Geography Indivisible

In a context of disciplinary division between human and physical geography, the book seeks to reassert the unity of the field through an emphasis on a shared focus on the geographic configuration of things and how and why configuration is important.

It first examines previous approaches to reestablishing unity, and why they have failed, before moving on to an explanation of fundamental differences in what is being studied and how. The role of configuration looms large in both. This is in the sense of contingency and the idea of emergence, suggesting that reconstruction of unity can proceed through an exchange of models of understanding.

This book will appeal to those teaching courses or seminars in geographic thought or in the history of geographic thought.

Kevin R. Cox is Emeritus Distinguished University Professor of Geography at Ohio State University. He is the author of numerous books, including *Making Human Geography*, *The Politics of Urban and Regional Development*, and the *American Exception*. He frequently blogs on his website Unfashionable Geographies.

Geography Indivisible

How and Why Configuration Matters

Kevin R. Cox

LONDON AND NEW YORK

First published 2023
by Routledge
4 Park Square, Milton Park, Abingdon, Oxon OX14 4RN

and by Routledge
605 Third Avenue, New York, NY 10158

Routledge is an imprint of the Taylor & Francis Group, an informa business

British Library Cataloguing-in-Publication Data
A catalogue record for this book is available from the British Library

ISBN: 978-1-032-42413-2 (hbk)
ISBN: 978-1-032-42415-6 (pbk)
ISBN: 978-1-003-36270-8 (ebk)

DOI: 10.4324/9781003362708

Typeset in Times New Roman
by Apex CoVantage, LLC

To the Taunton Three

Contents

Figures

Tables

Preface

The background to this short book is the major division within the field of geography between human and physical geography. It used to be that every academic geographer, from school teachers to university lecturers, had to have some background in both of them. At the university level, this is now very rare indeed. People specialize in one or the other, and then specialize even more. Physical geography is also divided, and to the extent that bits and pieces will lodge in different departments within the university: geomorphology in Geology or even in new, imperialistic sounding, Earth Sciences Departments. Climatology can find itself in a Department of Atmospheric Sciences. Famously, in Sweden, human and physical geography lodge in different departments, though still under the "geography" heading.

This fragmentation is a worry, and for several different reasons. There is a widespread popular interest in geography born of curiosity about places, but it is one that cuts across human and physical divisions; landscape is both cultural and physical. Ultimately journals like *National Geographic* and the *Geographical Magazine* need to be nourished by serious academic research. The bipolar nature of the field has also posed problems in how it is understood in academic circles and how or whether it can present and explain itself in a coherent manner. In the past, problems in its identity have contributed to the closure of university departments. So in terms of its external academic relations, it is important that we be clear on what the field is about and where its unity lies. Vague soporifics like geography as the study of the "earth as the home of man (*sic*)" will, reasonably enough, be suspect.

The major obstacle in the past to bringing the two fields together is that, in virtue of differences in their objects of study, any grand theory is impossible. How do you bring under the same theoretical framework people and rivers? There is another possibility, however, and that is to recognize the underlying general interest in geographic arrangement, how it occurs, and the effects that it has. This is the approach taken in this book. What the two fields share is an interest in the configuration of forces and things and how it makes a difference to understanding not just different places, but more general principles of understanding.

Many of those of us who did geography at the university level back in the 1950s and 1960s can recall the pleasure we experienced in studying both of the subfields: a pleasure born of the sheer aesthetic joy provided by geographic pattern, regardless of substance, and by the fact that it equipped us for a more comprehensive understanding of different places. Evidently the interest is still alive. A paper that I wrote back in 2006 on how one could bring physical geography into that staple of graduate seminars, the seminar on geographic thought – typically very biased toward human geography – is still widely read. According to ResearchGate, it had clocked up by March 2021 a quite massive 21,000 reads and still records about 150 per month.

This very brief contribution to the topic is divided into three chapters. The first sets out the way in which the two fields share a geographic imaginary suggesting that the division can be bridged, and then the various ways in which people have tried to do it, all of them wanting in various ways. The second chapter focuses on the divergences: how substance makes a difference in theorization, and correlatively, how the status of that key geographic concept, space relations, is understood differently. In the final chapter I set out an approach to bringing the two subfields closer to one another so that there might be, as the late Doreen Massey hoped, a conversation between them. This will be through the concept of configuration and what it entails for understanding the roles of contingency and complexity in both human and physical geography.

I owe a debt to three people who have read the manuscript and made encouraging and constructive comment on it. Many thanks, therefore, to Paul Claval, Nick Clifford, and Jonathan Phillips for their careful consideration. I am responsible for any shortcomings, but without their suggestions and questioning, there would have been many more. I would also like to acknowledge the work of Jim DeGrand, who prepared a number of the figures and diagrams.

Finally, I dedicate the book to "the Taunton three." The Taunton bit is purely symbolic and refers to a field course I attended back in 1959 that was based there, and where, one evening over dinner, as a humble undergraduate, I found myself sharing a table with three people who would have an extraordinary influence, in often quite different ways, on my thinking: Dick Chorley, Peter Haggett, and David Harvey. Dick inspired my interest in physical geography as a whole; I can still recall a lecture he gave on the subtropical high-pressure cells and their significance. Peter shared with Dick a desire to see change in the field as a whole and this book reflects those changes that they were so instrumental in bringing about. Peter was "the compleat geographer" with interests in both human and physical geography. He was also important in my early education as a spatial-quantitative geographer, something utterly foundational for my thinking, if latterly as counterpoint. David's influence has been equally different. He revolutionized how we study human geography and his geography has always shown through. I count myself fortunate in my academic life to have had such exemplars and mentors.

1 Unity in Diversity

"Geography is the art of the mappable."

– Peter Haggett

(https://peterhaggett.wordpress.com/in-his-words/)

Introduction

Peter Haggett is absolutely correct. But beyond that, things get much more complex. We can map cities, the weather map is a daily staple, mountain ranges, coastlines, and land uses. But the huge variation in the sorts of forces that produce these mappable entities challenges the idea that the field has a self-evident unity, a problem that has sometimes made its survival at the university level, particularly in the United States, something of a challenge. How to give coherence to such massive variation in objects of study? Fields of study are typically characterized and defended in terms of a distinct object of study, but geography seems to lack one.

The diversity of objects entering into geographic research has not stopped people looking for a unity beyond the merely mappable (Clifford 2002; Goudie 2017; Massey 1999b), and just to complicate things further, the mappable is something that geography shares with a number of other fields like epidemiology, geology, botany, ecology, archaeology, and city and regional planning: all what might be called spatial sciences and suggesting that the marriage of human and physical geography, the two major subfields of geography, might have had something of the whiff of the shotgun about it.

Human geography focuses on the way in which social processes express themselves over space and are in turn enabled and limited by those spatial arrangements. Physical geography is interested in the geography of the surface of the earth, including its biota, and the atmospheric envelope above it: how and why the land has the form it does; how and why vegetation varies

DOI: 10.4324/9781003362708-1

geographically in quite patterned ways; and then the "how" and "why" applied to the variation of the world's climates – to paraphrase the title of an old textbook on physical geography, *The Skin of the Earth.*[1] Unsurprisingly, perhaps, the two subfields have tended to diverge. This was not always so evident. But now they have their specialist journals, geographers specialize in one or the other, and finding a common language in which to converse is not easy. The assumption in this short book, though, is that there *is* a unity to the two, that there is a common language that can be exploited. The assumption has always been that this unity is not to be found in a particular object of study, though I will challenge that in my final chapter, but in a particular sort of imagination to be applied to a wide diversity of different materials, sometimes defensibly natural in origin, sometimes human, and at other times quite clearly of a hybrid nature; this assumes, of course, that people are both of the natural world and distinct from it.

In the remainder of this chapter I discuss how things used to be prior to the 1960s, and how there is a shared geographical imagination. There has been an invitation to try to integrate the two subfields. There have been several attempts but none has succeeded. And the reason for that, to be developed in the following chapter, is the rather obvious one: spatial form, aka geographical arrangement, cannot be understood without reference to process, and the processes undergirding human and physical geography turn out to be quite different, which in turn entails equally different methods.

Once Upon a Time

Where geography is a school subject, it remains the case that human and physical geography receive equal attention. In the United Kingdom, geography is one of the subjects examined in national exams at the ages of 15 and 17 approximately, and again, pupils are examined in both areas. This continues at university level though by the third and final year, students will be specializing. And if they continue beyond to an academic career in the field, that specialization will become very clearly marked.

It was not always so. Rather the changing social relations of the field have served to divide it. At one time geography as a field was inspired by a singular imagination, and specialization was limited. It was a field considerably detached from material concerns, even while it drew on fields like geology and economics that had much closer connections. In fact it was the ideal to be the "compleat geographer" and to work in both human and physical geography. Prior to the 1960s, a few academic geographers still addressed themes in both human and physical geography. This was particularly the case with those whose broader emphasis was on the former. Dudley Stamp is noted largely for his work in human geography but he was also active in the interpretation of physical landscapes (Stamp 1946); in fact, he had originally been a geologist. The American geographer Glen Trewartha made his mark largely in population geography, but he was responsible for one of the best climatology texts (1954) for undergraduates who were at university in the late 1950s. Yi-Fu Tuan is the doyen of humanistic geographers. But early on in his academic career, he studied pediments in arid climates (1959).[2]

Likewise human and physical geographers alike would do studies of particular regions, again working across the boundary. A book edited by J. B. Mitchell which appeared in 1962 under the title *Great Britain: Geographical Essays* consists of 30 chapters, of which 27 were devoted to particular regions. Each region was discussed from both a physical and a human geography standpoint. Each invariably began with a discussion of the lie of the land; notably how geology had affected the topography of the area, though the significance of that knowledge for what followed in the further discussion might seem quite remote. Physical geographers prepared chapters as did human geographers, each trying to work across respective divides, but, it has to be said, falling far short of any coherence (see Table 1.1).

The regional emphasis is possibly significant, suggesting that as the interest in regions and in regional geography declined with the growth of the spatial-quantitative work and its law-seeking emphasis, so did the interest in trying to bridge the human-physical gap.[3] This should be kept in mind when we turn to possible approaches to reintegration. At any rate, this happens no longer. All that remains is a sort of nostalgia genre where, in their later years, geographers cross the boundary to write regional geographies

Table 1.1 The incoherence of classical regional geography: examples from Mitchell (1962)

Chapter 8: The Upper Thames Basin
- Scarplands of the Upper Thames Basin
- Some Pleistocene landscape features of the Upper Thames Basin
- Oxford as a regional center
- The main road and rail network of the Upper Thames Basin

Chapter 10: Central South England
- Geological structure
- Geomorphological features
- Settlement pattern

Chapter 14: Rural Wales
- Principal erosion surfaces
- Density of population
- Settlement patterns
- Density of Welsh speakers in relation to total population

of the areas in which they grew up. Recent additions include Peter Haggett's (2012) book on the Quantocks and Allen Scott's (2014) on what he calls Solway country. Otherwise, the separation between the two branches has deepened to gulf-like proportions. One is either a human or a physical geographer, and that is that.

One can only speculate on why this might have happened. My suspicion is that the pressures have been both internal and external. There have been demands from outside the field for more specialized understandings, ones that can be turned to practical application. Harvey argued (1973: 124) that part of the context for the spatial-quantitative revolution in human geography was the growth of the planning field and, one would add, the increasing interest that human geographers showed in it. The journal *Regional Studies* was an early expression of this, and the first editor, significantly, was Peter Hall, who drew on other geographers for his editorial team. Physical geography has its own stories to tell. The early studies in hydraulic geometry were financed by the US Department of the Interior and for the very obvious reason that they were interested in rivers because they thought that it would help in understanding flooding. In climatology, as we will see, the air mass revolution reoriented climatology in the form of synoptic climatology toward more practical applications of a forecasting nature.

Geographers of both stripes, human and physical, have eagerly participated in these centrifugal effects. In part it was to escape the charge of amateurism that had been launched at geography, and particularly at human geography, in the immediate post-war period, a charge heavy with doubts about the future of Departments of Geography.[4] They have then found, though, that this fitted rather nicely with the competition for visibility in the field, something encouraged by Department Chairs eager to improve the field's standing in the university pecking order. Where careerism came from is unclear, aside from the struggle for promotion and tenure, but it is now a marked feature of university life and specialization, and even more specialization, offers the opportunity for carving out one's own niche, so long, that is, that one can connect it to broader ongoing research themes in the field. In the United Kingdom, the so-called Research Assessment Exercise has added its own impetus. Why care about the coherence of the field when someone somewhere is simply doing sums, but important ones?

The value of a more unitary geographic imagination, though, persists. It still excites imagination whether it is Bill Bryson dealing with the strangeness of the Brits, including their passion for the mountains; or Bruce Chatwin talking about Patagonia, the desert and Welsh settlement. Speaking personally, I had absolutely no interest in Australia until I read Bryson's *The Sunburnt Continent*. There is an intense popular interest in landscape in all its facets and in foreign places, again across all the elements of human and physical geography, as in the ongoing success of *National Geographic* or *The Geographical Magazine*. Academic geography has lost this more total interest.[5] A major goal of this book is to try to revive it.

A Shared Geographical Imagination

At the heart of geography is a curiosity about difference across the earth's surface: differences in climate, landforms, vegetation, human settlement, land use, degrees of urbanization. The geographer's autobiography typically takes off from an interest in, a curiosity about, difference – why the English Lake District looks different from the Pennines; why the Great Plains provide a different landscape from the so-called Basin and Range to the west; or why Northeast England had coalmines but East Anglia did not. And so on. As a result, analysis often starts with a map of some distribution: of national land use or population distribution or of land values in cities, or of the incidence of cirques or severe storms. This has then led to some simple classifications in both branches of the field with a view to capturing some crucial relations. So in human geography one might talk about geographically uneven development, centers and peripheries, settlement patterns, or different types of city, as in world cities. In physical geography there are discussions of particular sorts of landscape according to the dominant erosional controls – desert, fluvial, glaciated; or in climate studies, different sorts of climate: west coast temperate, Mediterranean, equatorial, temperate continental. In both again, there is a history of regional studies which are more concrete still and which characterize, or attempt to characterize, particular parts of the world in terms of a unity of conditions and outcomes even while they have tended to wane in favor of more general studies attempting to abstract from the complexity of the concrete.

Likewise, over the long history of the field there have been very similar approaches to taking difference across the earth's surface into account. The emphasis has been on space relations. To avoid misunderstanding, this requires explication. In human geography, the fact of the spatial-quantitative revolution or SQR of the 1960s (Cox 2014: Chapter 2) with its emphasis on the geometry of landscape has had some misleading associations. One result of its particular emphasis was to link "spatial" with what John Nyestuen (1963) called "fundamental spatial concepts": notably, distance, direction and connection. The SQR, in its search for laws of location, also explicitly positioned itself against an earlier, more particularity-inclined human geography. More implicit, though, was a rejection of an earlier human geography which focused more on areal association. Stress on relations between people and their so-called natural environment provide examples of this: so the great population voids of the world's deserts; or the way in which in the nineteenth century, the coalfields of Western Europe became major growth areas. This though is as spatial as central place theory. Rather the connections over space are more immediate so that it did indeed make sense to talk of the "environment"; that is, that which environs or surrounds.

Moreover, the idea of areal association has always been present in physical geography. The relation between topographic form and the resistance of different rocks to erosion, or that of scarplands to the tilting of sedimentary strata. Climates were linked to the annual march of the pressure belts as in Mediterranean climates oscillating between controls by desert high-pressure systems and the storms associated with the low-pressure belt to the north.

Moreover, there was always a sense of the spatial which hinted at the sort of spatiality brought to fruition in human geography in the SQR. Long before then, human geographers were writing about the significance of nodality, and of agglomeration. Mackinder's *Britain and the British Seas* was full of it. One also thinks of Köppen's well-known idealized continent showing the spatial relations of the different climatic types, or even the changing landscape ensembles envisaged by Davis's cycle of erosion, point in the same direction.

With regard to how geographic understanding has developed, there is something else. This is the quite fascinating way in which human geography and physical geography, or rather geomorphology and climatology, have tended to move in tandem, though not in any clearly coordinated way. I am referring here to the way in which differentiation across the earth's surface as a whole has been understood, and how this has changed. There have been quite radical shifts, culminating in the sense of wholeness and of interaction on a global scale that has finally reached all branches of the discipline. Geography's relation to the global has an intriguing history.

In a paper in 1991, Andrew Sayer tried to clarify some terms that had emerged in the context of a debate in human geography about locality studies. His concern was with the way in which the duality of general/specific was being drawn on. Generality, he argued, had two meanings. There was what he called "Generality I" and "Generality II." Generality I referred to widely replicated instances, often in form of a regularity or taxonomic group: so the different climatic types found in different locations around the world, or the world's wheat belts, the Great Plains, the Argentinian Pampas, the black earth of the Ukraine. Generality II, on the other hand, meant "large in relation to what one is looking at but perhaps internally related to it," as for instance in a spatial division of labor or the morphometric law connecting the length of a river to the area of its drainage basin.

Despite Mackinder and the classical geopolitics of early twentieth century (see Cox 2014: 242–247), it is the first sense of the global that dominated geography, both human and physical, at least until mid-century. Geomorphology long recognized the significance of variation over the earth's surface as a whole. Davis's different climatically controlled cycles come to mind along with the more focused climatic geomorphology. Wegener's continental drift had been suggestive of Sayer's Generality II, but it was the more broadly accepted theory of tectonic plates that laid the foundation for what Mike Summerfield would call "global geomorphology" (2014) proposing internal relations between landform complexes and the movements of the earth's continents generating uplifts, mounting building, and rift-formation.

In very broad terms, and as F. K. Hare laid out in his (1953) book *The Restless Atmosphere*, climatology underwent a similar revolution. His emphasis was on what he called "dynamic climatology" which focused on the global circulation of the atmosphere as a means of making sense of the different climatic types. In his Preface he distanced this approach from what had hitherto passed in geography for climatology. In the hands of people like Köppen, Kendrew (1922) and DeCourcy Ward (1918), this had been static, trying to explain regional climates in terms of the pressure belts, receipts of solar radiation, the presence of ocean currents, continentality, the distribution of relief and such like. His book put its focus elsewhere: on the movement of air masses on a global scale and the global circulation of the atmosphere which controlled those movements.[6]

Human geography has undergone a similar evolution. A landmark study, at least in the history of British geographic thought, was Herbertson's (1905) paper on "The Major Natural Regions: An Essay in Systematic Geography." In defining these, he drew mainly on climate and, to a lesser degree, relief. The goal, at a time when geography was seen as the relation between people and their natural environment, was to shed light on patterns of the same: that is, one could hold the natural environment constant and see what people made of it, and why the differences. This sort of approach was the basis of Ellsworth Huntington's work on what he called "civilization and climate." We should not, though, be distracted by the racist interpretations that were given. Herbertson's title referred to "systematic geography" and this would be a continuing theme all the way to the 1950s. Something like John Alexander's (1963) *Economic Geography* looked at distributions on a world scale but from the standpoint of Sayer's Generality I: similarities and differences across the earth's surface in economic activity and what might account for it in terms of other, co-varying conditions.

What put paid to that approach was change in the material world: the rise of what would become known as "globalization." Peter Taylor's paper of 1981 clearly reflected this, referring to a scalar division of labor framed by exchange at a global level. The rest, all the work on globalization, the local-global relation, is history. Sayer's Generality II had arrived: human geography had converged in its understanding of the world and how it framed areal differentiation, alongside geomorphology and climatology and even though the histories of those trajectories were quite different.

Yet while human and physical geography share a similar imagination, you might not be aware of it. Human geography is extraordinarily aware of its basis in the study of space relations, while physical geography is utterly and completely unselfconscious about it, for reasons to be explored.[7] Nevertheless, both draw on John Nyestuen's (1963) fundamental spatial concepts of distance, direction and connection and derivative ideas of scale, shape, area, pattern and landscape. Human geography's spatial order in the form of segregation, urban spacing, urban densities, industrial clustering, or whatever find a counterpart in the landform ensembles of glaciated or fluvial landscapes, in the repetitive space relations of climates on Köppen's idealized continent, or in Alfred Wallace's biogeographic regions. Directionality is as apparent in the movement of air masses as in the movement of people, or Sauer's agricultural dispersals. The connections between places in terms of specialization and trade find an analogue in climatology's teleconnections. Geomorphologists (Chorley et al. 1957; Das et al. 2022) have been as interested in shape as human geographers (Boyce and Clark 1963; Taylor 1971). And so on. This, though, is very static and to move toward a more explanatory position we need to identify analogous dynamics.

In particular, in both human and physical geography, as in the other spatial sciences, we deal with movements and fixities of various sorts. In geomorphology the focus is on the movement of materials: their weathering and/or erosion in one location, their subsequent transportation by water, ice or wind, and their deposition elsewhere, as on floodplains or in deltas. There is a fixity to the physical landscape: when we look at it a year later, it will be seemingly identical to what it was when we first saw it. But there are also movements underway, albeit slow ones; movements which will, over millions of years, transform it. Climatology is also about movements, notably those of the air and of the ocean; one talks about the general atmospheric circulation, and the role of ocean currents. But again, there are also relatively stable features to world climatology, as in the relative permanence of the subtropical high-pressure cells north and south of the equator. These "march"[8] northwards with the northern summer and then retreat southwards in the northern winter but their average position remains virtually the same.

A more graphic example might be provided by the hydrological cycle. Water is stored in various places: lakes, ice sheets, oceans, and underground in aquifers. It also moves around: Evaporation converts the water of the oceans into water vapor. This is then transferred landward by winds in the form of so-called air masses. Once over land, and to the extent that the air mass is unstable, the air rises and the water vapor condenses as rain. It then moves back to earth where it enters streams which in turn enter lakes, etc., and eventually reaches the sea; or the water is evaporated from stream and lake surfaces; or it falls over northern latitudes as snow and becomes part of ice caps.[9] And so on: movement and fixity, describing a geography that we can map.

We can consider human geography from a similar standpoint. People move around and they settle in particular places. These places have histories of growth and decline; more people move in or move out so that the settlement pattern changes over time. This is to think in quite ahistoric terms, which has its limits when we talk about objects which revolutionize their social relations and so their spatial practices. A better example, therefore, is the circulation of capital in capitalist societies. Capital describes a circuit. It assumes mobile forms as in the case of the money laid out for production, the raw-materials moving from mine to factory, or the wage workers moving from home to places of work. It also assumes more fixed forms, notably factories, mines, and farms but also cities. Without those movements none of these fixed forms could continue to exist. Think likewise of the making of the modern world: the vast movements of people and capital from Europe to the neo-Europes of North America, South America and the Antipodes and the creation there of the relatively fixed physical forms of railway networks and the social forms of nation states.

However, nothing in the world is fixed in an absolute sense. Even the continents "drift." Without the movements that sustain them, fixity dissolves only to be reconstructed elsewhere in virtue of changing movements of capital, of people, of water as people construct reservoirs, for example, and of the landmasses with their implications for frontogenesis and therefore for storm generation. Harvey (1996: Chapter 2) has suggested, therefore, that in talking about fixities, we use the term "permanences," retaining the scare quotes. Nothing is truly permanent. Unless cities are maintained through the application of capital to their physical fabric, then they disappear; so-called ghost towns are just one stage in that process of disappearance, though they might find themselves preserved in a tourist aspic.

Finally, discussion of the cycles raises some interesting questions. Some have seen it as a way of integrating human and physical geography.[10] In virtue of their physiological needs, people are part of the hydrological cycle, just as they are parts of the nitrogen and carbon cycles. People also are important climatic and geomorphological agents and increasingly so. It is the "increasingly so" that is the worrying bit and which puts severe limits on such an integration. This is because it derives from the fact that while people are part of nature they are also in virtue of their social "nature," outside of it. Only people have industrial revolutions. To investigate the people-nature relation as a unity, therefore, means stepping outside the realm of natural law and that means drawing on an entirely different body of concepts and theories.

Attempts at Integration

Geographers, particularly British geographers, and those in the former empire trained in the British tradition, have worried about this division. There have then been attempts to bring the two branches together, though with a success that, in my view, should be qualified. They are either one-sided – an integration from a position in human geography rather than one in physical geography, or vice versa – or based purely on formal considerations of an analogical nature. Whether it can be more than this remains to be considered.

Traditional People-Environment Studies

For much of the first half of the twentieth century geography was the study of the relationship of people to the so-called "natural" environment. The copresence of human and physical geography under the same roof was then justified in terms of a shared interest in the natural environment. How much was shared, though, was always in doubt. Just how much of physical geography did a human geographer need to know in order to make sense of the relation of people to the physical environment? For a human geographer interested in the American Cotton Belt, it was important to know that the northern limit of cotton cultivation was 200 frost-free days, but an explanation of the geography of climate in the United States was not. Likewise, one could point to odd facts like the location of schools on river terraces where they could take advantage of flat land for playing fields without knowing exactly why river terraces existed. A human geographer might draw on descriptions of the physical environment therefore, without having to understand how the objects to which they referred actually came about. For a physical geographer the relation was even weaker. Physical geography was studied as if people did not exist, which is in some and increasing contrast to physical geography today.

This problem, though, needs to be set in a broader context. It is not just that human geography can draw on physical geography without entering into the reasons for a particular climate or lie of the land. As discussed earlier, physical geography itself has its own distinct cleavages of an analogous sort. How much climatology does one need to investigate land form geography? In examining desert landscapes how important is it to know *why* rain, when it falls, tends to be extremely intense, with implications for erosion? The same applies in reverse. The disposition of the landmasses certainly affects climate but a knowledge of continental drift is not important, unless, that is, one is engaged in reconstructing past climates.[11] The rise of fields like ecohydrology, critical zone science, earth system science, and others claiming a more holistic approach to the physical world does not qualify this conclusion: contingency relating climatic processes and those governing landforms continues. Landform development cannot be reduced to climate change and so on.

"Man's Role in Changing the Face of the Earth"

Something of a landmark in the history of geographic study was the publication in 1956 of a book with the title *Man's Role in Changing the Face of the Earth,* and edited by William L Thomas (1956). Putting to one side for the moment the genderized title, one significant aspect of the book was, indeed, as a contribution to what I defined earlier as the study of people in their relation to the environment. Over the course of the first half of the twentieth century, the latter had visualized the so-called physical environment as, at worst from the standpoint of its defensibility, acting on people; and at best as a passive condition of their lives, or something to which they had to adapt. In contrast, *Man's Role* focused on the way in which the physical environment was something to whose formation people contributed, sometimes deliberately, as in clearing the forests; and sometimes unintentionally, as that same deforestation increased runoff and so enhanced the erosive power of streams. In some of the chapters, on the other hand, there was more attention to the interplay of people and "the face of the earth": how people were a force in changing the land and its vegetation but, in a very few cases, how those particular interventions might then mandate some shift in the way in which they organized their social relations.

The human role in changing the environment has been an ongoing theme in geography ever since. There have been books on the topic (e.g., Goudie 1981), some of them weighing the pros and cons of human intervention in particular instances, and numerous articles not just in geomorphology (e.g., Graf 1979) but also in climatology, as in the work on urban heat islands. One might also single out the book by Alfred Grove and Oliver Rackham (2003), both for its classic debunking character and its grasp of the processes – physical and human – at work, and over a very long period of time at that.[12]

The literature taken as a whole, though, is intellectually unsatisfying. One reason is that it lacks theory. It is a highly empirical literature. Of course one needs theory to understand how an urban heat island functions or how it was that Medieval peat diggings were flooded to create the Norfolk Broads (Lambert et al. 1960). But exactly *why* human intervention to begin with? How does one situate it, its changing forms, and the intensity of its consequences with respect to some idea of social process? And how do concrete studies of this sort feedback to help in reformulating theory?

Social process is a glaring absence in this work, something that has been partly addressed by critical physical geography, to be reviewed below. Moreover, it is not just that people change physical geography; their changes to the world, the sorts of practices that those interventions entail, also have consequences for social organization. The controversies surrounding Wittfogel's ideas about hydraulic society notwithstanding, as

Harvey has emphasized, ecological projects are also social projects[13]; they imply and then entail certain sorts of social relation.

The Spatial-Quantitative Revolution

The spatial-quantitative revolution opened a new window on the possibilities of convergence. This stemmed partly from work on spatial order, something that emerged quite by chance at very similar times, both in urban geography and in geomorphology: the order of a central place landscape seemed to be matched by the order in fluvial landscapes revealed by the laws of morphometry. And also the language of statistics employed in both was not just something shared: it also seemed to offer a way out of the charge of amateurism that had afflicted both human geography and geomorphology. Henceforth geography would be "scientific."

The high priest of this gospel was William Bunge (1962). He believed that the unity of the field could be reestablished by a focus on space relations. His book is notable for the way it abstracts from any notions of substantive process. Movements could be anything: migration, advancing ice sheets, the diffusion of ideas, the circulation of the atmosphere. The same applied to patterns: the pattern formed by rivers in a drainage basin, the distribution of cities, of climatic zones, or whatever. He believed that his theoretical geography applied regardless; it could therefore embrace both human and physical geography and so resolve the division in the field.

What he called the nearness problem was manifest in both: air masses move from high pressure to the nearest low pressure, albeit structured in their directionality by the Coriolis force subsequent to the rotation of the earth; in economic geography there is a net migration from lower wage areas to higher wage areas, but so as to minimize movement (1962: 211). Likewise in contemplating a uniform distribution of points along a line:

> One method of obtaining a grasp of the power of the pattern is to stare at the unlabeled pattern and ask yourself "Of what is this a map?" Some possible answers include filling stations along a highway, major volcanic peaks along the Cascades and the distribution of ice cream vendors along a beach. Note that these suggested applications to the earth's surface are more than shallow spatial coincidences. For instance the total travel cost along a beach for the consumer of ice cream is minimized by such a pattern. The volcanic pattern minimizes the movement of magma in the fissure, or put in another way, the uniform distribution marks points of the greatest internal pressure.
>
> (1962: 254)

In other words: Highly imaginative and stimulating. Even so, the insistence on location suggested a human geography bias in his understanding and, as it turned out, his assumptions about the world and those of spatial-quantitative geography had limited effect on physical geography.

There were some exceptions. The British geomorphologist Richard Chorley was to the forefront and he and Peter Haggett co-wrote a book on network models in human geography (1969) that was intended to bridge the divide. Peter Haggett (1967) wrote a highly imaginative piece which tried to transfer Horton's notions of regularity in the geometry of drainage basins to transportation networks. Trend surface analysis was drawn on in both physical and human geography. Michael Woldenberg (1969) then drew on analogies between central place geometry and the spatial order apparent in drainage basins.[14] There was also some interest at the University of Iowa: a major center of spatial-quantitative work at that time – topics like the geomorphic significance of the clustering of sink holes in karst topography. Climatic geomorphology looked at the relations between climate on the one hand and particular landforms on the other. But the impact on physical geography overall was unremarkable. This was aside from some methodological contributions like those of Rayner and Golledge (1972) on applying spectral analysis to human geography, while Rayner's earlier work (1971) had been its use in climatology. The fact was, as Robert Sack (1972) argued, there could be no science of space relations. Substance made a difference. Any attempt at bringing human and physical geography together has to put that in the foreground.

Landscape

There is a strong tradition in geography of landscape study. This is so in human and physical geography alike. In physical geography, it is something going back to the turn of the last century – Marr (1900) and Lubbock (1902; aka Lord Avebury) and then picked up later with books like Arthur Trueman's *Geology and Scenery in England and Wales*, Dudley Stamp's *Britain's Structure and Scenery,* and Wooldridge's *The Weald.* Later there would be Bruce Sparks' *Rocks and Relief.* Even climatology has had a look in, most notably in Gordon Manley's *Climate and the British Scene.* Manley described not just how the shape of the land affected the country's climate but also how the changing light brought about by the seasons, and then the weather, had effects on how we perceived the landscape. Landscape has also been a theme not just with geographers, as in Darby's (1951) essay on the changing English landscape but also economic historians, as in W. G. Hoskins' (1955) highly regarded *The Making of the English Landscape* or Maurice Beresford's (1954) on deserted villages. It was central to the Sauerian school of American geography.

Bringing physical and human geography together through landscape studies has been more elusive, and has largely occurred through the role of human intervention into natural processes, as in the work on the clearing of the woodland or the draining of the Fens: both research foci of the historical geographer H. C. Darby, but with the same weaknesses as the literature on "man's role in changing the face of the earth." Nevertheless, the fact of the interest in landscape is significant. It is a popular one, and a major way in which the lay person makes contact with geography. It also testifies to geography as a field in which the visual is of major importance.[15]

One of the reasons for this is aesthetic. The first chapter of John Marr's *Scientific Study of Scenery* (1900) opens with the words,

> A widespread appreciation of the beauties of nature is not the least of the many beneficent changes which have marked the Victorian era, and with this appreciation has sprung up a desire on the part of many people to obtain some insight into the causes of scenery.

This would be echoed over a century later by Andrew Goudie: "The main stimulus to an interest in geomorphology is an interest in visually appealing landscapes. One can often detect in the writings of some of the pioneers of geomorphology a clear enthusiasm for understanding great landscapes" (Goudie 2002: 245; see also Dixon et al. 2013).[16] The aesthetic is also a theme in Gordon Manley's book; many of the numerous photographs have clearly been included with this in mind.

This, though, suggests a weakness in landscape studies. The emphasis on a particular version of the visual, one which brings into focus order in the landscape, results in an undynamic approach to geography. This is the case whether or not the relation between people and the environment is of interest. Current landscapes tend to get interpreted as an adjustment of people to nature and a harmonious adjustment at that: castles atop bluffs, building materials reflecting the local geology, rivers artfully framed by bridges.[17] This is something captured by coffee table books, suggesting the deep human or even social resonance of this sort of relationship: A reflection of alienated times in which harmony in the landscape serves as a compensation for a harmony impossible in everyday life.

It is, then, a rather static geography that is being portrayed: one devoid of the signs of tension and contradiction latent with transformation.[18] This is the case even while many physical landscapes are in a postglacial disequilibrium state: something masked by human adaptations and the slowness with which the landscape returns to an equilibrium between form and contemporary process. This is a problem not specific to landscape studies. The spatial order sought by the spatial-quantitative geographers, as in the nice equilibrium distributions of central place theory is an important case in point. All of which suggests that in any integration of physical and human geography it is time-space that should be our point of reference and not in a mere descriptive sense. Rather it is how change is the normal: something implicitly denied by most of the landscape geographers.

"Historical, Complex Sciences"

In 1999, in an attempt, as she put it, to get a conversation going between human and physical geographers, Doreen Massey published a paper on what they shared. Her conclusion was that they were both historical and complex sciences. Taking each of these claims separately, she meant first that both drew, or should draw, on a conception of the world in the three dimensions of time and space, as indeed was the case in some well-known instances. She did not mention Hägerstrand, but his work on diffusion, migration, and then what he called time-geography is, in that regard, classic. Likewise in geomorphology, there is a long history of concern with the development of the physical landscape in which events in space and time play a major role. From the 1960s on, this was eclipsed by a greater focus on process studies, but there are now signs of a revival of interest in long-term development. By complexity, she wanted to draw attention to the way in which complexity theory could be drawn on in both instances to illuminate respective geographies. This has made an appearance in the now rather clichéd idea of "path-dependent development" but there are earlier instances of its relevance, as in theories of agglomeration or of neighborhood effects or indeed the work of Hägerstrand. A fundamental idea here is the way in which chaos can give way to order: a particular juxtaposition of chance events can be the condition for relations of a nonlinear sort so as to produce something approximating regularity: so uneven development or the way in which rivers can cut across scarplands seemingly regardless of variations in the resistance of the underlying geology. The way in which history and complexity come together is then clear. The development of a geography is rooted in the past and how past events in time and space interacted, and continue to interact, with one another. Otherwise put, both human and physical geography can be instances of storytelling in which the conclusion is entirely unpredictable, subject as it is to space-time conjunctures of an unforeseeable nature.

There are some issues here. In her account, the contingent seems to be the precursor of interactions of a complex sort, but that does not always happen. In her attempt to banish what she called "physics envy," and the pursuit of a more quantitative, predictive geography she risks voiding order. This is to miss the point that the historical and complex are themselves productive of order. In understanding the international division of labor, contingency and complexity are of the essence. The world's climates have a geography that is clearly ordered, but the details of that ordering are then subject to the contingent and the complex as global warming is demonstrating. Regardless, after an initial flurry of interest in what she had so say, attention quickly lapsed. This seems to me to be a missed opportunity and one that I will take up later. Any attempt to recreate the indivisibility of the field has to incorporate the ideas of time-space and complexity that she was trying to propagate.

Critical Physical Geography

More recently another attempt to bridge the gap between physical and human geography has emerged. This is critical physical geography, inspired by the approach of critical human geography. According to Rebecca Lave, one of its principal protagonists, this combines "critical attention to relations of social power with deep knowledge of a particular field of biophysical science or technology in the service of social and environmental transformation" (Lave et al. 2014: 2). Science, including physical geography, is not value free. Rather it has objectives and incorporates frameworks of understanding and method strongly influenced by prevailing social needs; and some people's needs, in virtue of the power that they wield, are more important than those of others. Physical geography as practiced, works for some and not for everybody.[19] The physical landscape is a hybrid: to understand it we need a knowledge not just of biophysical processes but also of the power relations that have had contingent or even necessary effects on how those processes get channeled.[20] The recent suggestion of a new geological period, the Anthropocene, in which human effects on physical geography have become accentuated, has added impetus to this approach.

The way in which this works is not necessarily obvious. Rather, and more typically, it is an outcome of a social process *in toto*, but one which works from the standpoint of those equipped with the power of money. The idea that the colonized got things wrong in their land use practices, that practice was a poor adaptation to the climate, wasn't necessarily because of a direct need on the part of the imperial powers to dominate and to justify imperialism, even while that need was present. Rather it flowed from a more general, racialized understanding of the indigenous populations as inevitably technically ill-informed, subject to prejudice and all manner of irrationalities, and in need of help. A failure to recognize the wisdom of customary practice, based on a long history of trial and error, would then often result in a degradation of the landscape and in the case of semi-arid environments, desertification.

Dominant, positivistic notions of science have also been part of the mix. As Brian Fay (1975) has pointed out, positivism was attractive to those bent on the exploitation of nature because it offered predictability and hence control. The ideology of expertise has given further impetus to this sort of work: how could those who have been subjected to years of education and training be wrong? Yet in environments characterized by extreme events, as in, again, semiarid environments, it has proved to be a weak basis for protecting the subsistence base of indigenous peoples or even the business prospects of ranchers.

There are some interesting precursors of this sort of approach as well as some more recent exemplars. Of the former, the work of Mike Davis is of particular interest. In a classic paper (1995), "The Case for Letting Malibu Burn," he shows how a particular configuration of physical geography and social forces has resulted in devastating fires along the Malibu coast, particularly during the autumn when the Santa Ana winds blow, often with ferocity, and always desiccating. Their effect is intensified by the north-south alignment of canyons debouching on the coast, channeling the wind: so a particular physical geographic configuration, including high pressure over the Great Basin which results in the northeasterly flow into Southern California. Fire is a natural, lightning-induced occurrence. Human intervention, structured by powerful social forces, has made fires less frequent but, in virtue of that fact, and the accumulation of dead, flammable vegetable matter, has made them far more intense and devastating. To start with, developers built houses up against the fire-vulnerable chaparral, in areas where they should never have been built in the first place: but the views were good and housing could be sold at high prices. The people who then moved in are wealthy, anxious to protect their property values, and with the power to do so. The expert view is that these massive conflagrations can be avoided by periodic, controlled, burns, which would impede the accumulation of fuel loads. But there is always the small risk of them getting out of control, and this has generated strong local opposition: local fire departments have retreated. Meanwhile, property owner obstinacy is enhanced by the way in which it is the general public which pays to put the fires out when they occur.

More recently Sayre (2010) has made some interesting and new connections between global warming and human intervention. It is not just human consumption of fossil fuels which is the culprit but particular forms of built environment which encourage it, most notably urban sprawl, which reaches its peak in the United States. As Walker and Large (1975) have pointed out, this has origins in an accumulation process promoted and defended by an alliance of the auto companies, the oil companies and real estate interests, including those of US banks. Their profits depend on the fact of an urban form that, in its implications for human movement, is energy intensive.

This also reminds of the disaster critiques of the mid-70s and onwards. The object here was the idea of the "natural" disaster. Vulnerability to periodic droughts in the Sahel region had been increased under the French Empire as a result of colonial policy, the way it interfered with agricultural practice, and the maintenance of buffer stocks for times of need (Watts 1983). The idea of "disaster" makes periods of drought seem like an accident, but as Hewitt (1983) observed, they are naturally recurring events, even while with global warming, one might add, periodicities might be changing, but historically, much more predictable than social change. The response has been primarily technical. This itself is embedded in a social system where money can be made from such interventions and where the social changes necessary to reduce vulnerability are ones that would infringe on those privileged by the status quo.

There are some limits to this as a way of integrating human and physical geography. There is a clear sense in which human geography *has* to be critical; this is not the case in physical geography. Social laws are power-infused but physical laws are not. There are, in consequence, huge chunks of physical geography that remain out of the reach of critical physical geography, including relict landscapes, most obviously but by no means exclusively, glaciated; processes that have no clear relation to human intervention like continental drift. Climate, of course, is something else, though the laws governing the balance between erosion and deposition will survive subsequent changes in stream volume and oceanic encroachment; and the hydrological cycle will continue. And while indeed political priorities might result in bad physical geography that is misleading from a policy standpoint, it does not have to be thus. The work of Leopold et al. (1964) on rivers was presumably funded because of the light it might shed on flooding and hence be of interest to, say, insurance companies. But that does not make it bad science.

Summary Comments

The term "geography" when applied to what human and physical geographers, respectively, do is entirely apposite. At a high level of abstraction, they share similar fascinations with the way things are distributed over the face of the earth, with landscape, with geographic scale, questions of nearness, and with other things like the relation between fixity and movement of which they might be less aware, at least when stated thus. But while human geographers – economic, cultural, urban, political – can have useful conversations with one another, getting something similar going between human geographers as a group and physical geographers taken together has been much more difficult. They go to different sorts of conferences, often interdisciplinary, and they have their own research outlets.

The news gets even worse, though, when one recognizes that while holding human geography together is not too much of a challenge, doing the same for physical geography really is: what *do* geomorphologists, biogeographers, and climatologists share, other than, that is, their geographical perspective on the world and a dedication to processes, albeit different ones, of a physical nature? The knowledge of climate needed to study geomorphology is very limited. There is something called climatic geomorphology: how landforms vary according to the dominant climate, but a detailed knowledge of climatology and its processes is superfluous. And, aside from the importance of continental drift for the distribution of land and water, and therefore for climate change, climatologists can proceed in complete ignorance of geomorphological process. The same applies to biogeography: the facts of climate are crucial but not the processes demarcating different climatic regimes. In contrast, human geography has the clear potential to be integrated, to the extent that one opts for some unified, internally consistent, concept of the social process: a Marxist or a Weberian human geography, perhaps. But the physical laws drawn on in geomorphology are different from those used in climatology, and biogeography also has its own laws, this time of a biophysical nature.

Nevertheless, these separations have not stopped attempts to bring human and physical geography together under an umbrella that can counter the centrifugal forces. By and large these attempts have not been as successful as one would have hoped. Quite what it is that makes the task so difficult is addressed in the next chapter. Few of course will lay awake at night worrying about it. This may be because the problems seem so insoluble. The underlying premise of this short book, though, is that there are more similarities than have been explored hitherto.

Notes

1 A Austin Miller (1953).
2 There are numerous other instances. Alfred Grove is noted for his work on desert landforms but he has also done work in human geography (1969, 2003).[2] John Borchert was noted for the most part for his contributions to urban geography, but earlier in his career he had made a more than respectable contribution to regional climatology (1950). F. H. W. Green is best known for his early studies of urban spheres of influence, but his major field of concentration was climatology.Other cases include Monica Cole, a biogeographer, who also wrote on the human geography of South Africa; Griffith Taylor, the Australian geographer, noted primarily for his neo-environmentalism but also with contributions to geomorphology; and John Leighly, noted more for his physical geography, also did work in urban geography. It was also evident in classroom teaching.

As an undergraduate I recall a human geographer of very considerable repute, and deservedly so, teaching a course on the regional geomorphology of Wales.

3 I am indebted to Nick Clifford for this suggestion.

4 Though no-one raised the same sort of complaint about English studies, History, or Modern Languages, where the claim might have applied equally well.

5 There are some notable exceptions; this from Doreen Massey (2009: 403): "But the other thing, which one can't live up to, but I think is a geographical potential, is the way geography crosses, or has the potential to cross, the human sciences and the natural sciences. And before I had to specialise, that was one of the things that I really, really liked about geography. The reason I am wearing the heaviest clothes I've got, including hiking boots, is because I am just off to Sutherland (I was there last year as well) and I'll spend three days remembering all the stuff about the Moine Thrust and be utterly gripped by geology. So it still has that potential absolutely to engage me, that 'holistic' perspective of geography."

6 On the timing of this "globalization" of climatology, compare Chorley: "Undoubtedly the most important outcome of work in the second half of the twentieth century was the recognition of the existence of the global climate system" (Barry and Chorley 2003: 6).

7 Some might object to my emphasis on space relations and claim people-environment relations as the core of the field. Aside from the fact that this applies in an unbalanced way to the two fields – physical geography exists regardless of the presence of people and will endure once they have vanished – "environment" is as spatial as it gets.

8 The imagery of "marching" is intriguing, conveying the necessary, determined character of what is happening.

9 I do not mean to preclude the fact that water vapor condenses and falls as rain over the sea. My example is chosen to illustrate what happens over land and how water circulates back into the atmosphere by various routes.

10 See Margaret Anderson's (1951) little classic *The Geography of Living Things*.

11 Goudie and Viles (2010: 32) note the implications of plate tectonics and its acceptance among researchers for an integrated earth science since it helped explain the long-term development of climates, ocean currents, vegetation and landscapes across the world. But one must note that the effects of continental drift on climate are of an entirely contingent sort and do not affect the basic laws governing climate.

12 Their focus is the vegetation of the Mediterranean and the view that it has been degraded in the course of human history and now in certain parts faces a future of desertification. This is the "ruined landscape" thesis: the degradation by humans of vegetation from once prolific woodland to a mix of forest, xerophytic brush and steppe. Accordingly it is human-induced fire and soil erosion that are responsible. Is this, however, true? And if not, how did the argument arise? What they show is that overall and with some localized exceptions the vegetation of the Mediterranean basin has changed very little over the last 4,000 years. Rates of erosion are relatively high but this has more to do with the unstable tectonics of the area than to deforestation. Likewise the thesis that vegetation has been degraded by human-induced fire is shown to be untenable. Naturally induced fires would have been common through lightning strikes long before people arrived on the scene. See Grove and Rackham 2003.

13 "Created ecosystems tend to both reflect and instantiate, therefore, the social systems that gave rise to them . . ." (1996: 185.) For an excellent case study that illustrates exactly Harvey's point, see Swyngedouw (2007.)
14 Thanks to Jonathan Phillips for pointing this out.
15 We should note, though, that in physical geography, the idea of landscape has been undergoing some mutation: less something visual to be explained, and more a coherent unit of biophysical processes: so a marshland as a landscape – certainly something visually distinct, but also as something for study quite independently. I am grateful to Jonathan Phillips for pointing this out.
16 Compare Peter Haggett (1990: 1–2) and how his interest was stimulated by Charles Cotton's diagrams of New Zealand landforms.
17 As in John Sell Cotman's wonderful "Greta Bridge."
18 Compare Dennis Cosgrove: "From a critical perspective, the pictorial dimension of landscape has frequently been charged with duplicity. Dissecting landscape's capacity to 'naturalize' social or environmental inequities through an aesthetics of visual harmony, geographers and art historians have long recognized that 'Georgian' landscapes, superficially paradigms of English social and environmental order, were often painstakingly constructed by rapacious landowners in the course of destroying more communal but less profitable fields, farms and dwellings" (2006: 51). Don Mitchell's approach, as in his *The Lie of the Land* (1996), is very similar, underlining the way in which landscapes express social relations of a highly contradictory nature.
19 To call for a physical geography that is socially "critical" should not overlook the fact that there is also a sense in which it is always critical: contesting accepted ideas as to the fundamental processes. It also works in the reverse direction. The more restricted sense of "critical" can also be applied by physical geographers toward the work of human geographers. Given their concern for climate change, one might be surprised that they have not been more active in calling for a critical human geography more sensitive to their contributions.
20 ". . . we cannot rely on explanations grounded in physical or critical human geography alone because socio-biophysical landscapes are as much the product of unequal power relations, histories of colonialism, and racial and gender disparities as they are of hydrology, ecology, and climate change" (Lave et al. 2014: 3).

2 Divergences

Human and physical geography are both, undeniably, "geography." For a long time, university degrees in the field required some competence in both. Some professional geographers practiced both. Nevertheless, the subdisciplines have moved apart and people have struggled to establish a basis for unity of a research sort between them: how to do research where the boundaries between human and physical geography melt away. For there are indeed boundaries. These are primarily ones of substance which then have implications for theory and for method. We start with matters of substance.

Making Sense of the Differences

A major contrast is in the nature of the fundamental processes and the objects that constitute them: social in the one instance and entirely natural in the other. Right at the outset some may object to this sharp separation. Obviously there is a mutual interpenetration of people and nature. It is in virtue of people's physical nature that they have been able to develop their distinctive social abilities, not least the capacity for highly developed systems of symbolic communication. Likewise nature is deeply affected by human intervention. However, it is not clear that through that intervention, natural processes are changed qualitatively. They are certainly changed *quantitatively*, including the achievement and surpassing of important thresholds that can indeed alter their qualitative nature. But human intervention is beside the point since we know that it is not necessary to unleash these changes. They can happen for entirely different reasons that have nothing to do with what people are doing or have done, and such changes were occurring long before the appearance of human beings on the face of the earth.

In human geography it is social processes that are to the fore. This has several consequences. Society "develops"[1] and with it, the human geography that we study necessarily changes. Hitherto at least, that change has assumed a clear directionality: not least, urbanization on the one hand and

DOI: 10.4324/9781003362708-2

greater interdependence over space on the other. The reason it "develops" has to do with human agency. Without the ability of human beings to conceive of different ways of doing things and to put those ideas into practice[2] there would be no institutional change, no technical change and therefore no change in the broad outlines of concrete human geographies; except those they would come to share with all other species as a result of processes of evolution. Social relations are inevitably spatial relations. As society changes so do spatial practices. People, in virtue of their social relations, acquire incentives to exclude, to change their locations in a purposive manner. It is reasonable to talk about human beings "locating" or engaging in territorial processes of segregation.

Physical geography is different. The objects of physical geography do not "act" in the sense that people do. They exercise no locational choice. They do not fight in a purposeful sense one with another over location; they simply behave according to their nature, which means that the laws governing them do not change. A drainage basin can encroach on another through processes of river capture, just as a cold air mass can undercut a warm one, but there is no intent involved. Air will flow from areas of high to ones of low pressure. A particular physical stimulus like lowering a floodplain through excessive pumping of underground water and resultant drying out and contraction of the particles making up the sediment, will, given a particular discharge of water down the river, result in floods of increased depth and extent. The undercutting of cliffs by rivers will eventually lead to their collapse. And so on. Space is part of, a fundamental aspect of, these physical processes and not separable in the sense that it appears to be in social processes, where people seemingly make choices, abandoning one set of locations in favor of another as social circumstances, including technologies, change.

Rather, and at best, in physical geography, one can refer to cyclical processes. There are short-term cycles like the annual movements of the climatic zones or the hydrological cycle; and longer-term ones like the – highly contested – cycle of erosion. But it is "at best": When it comes to discerning some order in their magnitude or change in their spatial relations, continental drift, climatic change, and tectonic movements remain elusive and conform more to a random walk.

It follows that scale relations are also different. In human geography one talks significantly of the "construction" of geographic scale. Prior to the voyages of discovery from 1492 on, it would have made little sense to talk of the world as, in social terms, something integrated on a global scale. Today it is more feasible, even while there are different degrees of integration depending on where one is located, both physically and socially, in the world. Yet the atmospheric circulation has *always* been global, likewise the circulation of the oceans: the pattern of currents may have changed but the water of the oceans has always been displaced around the world from one "ocean" to another. The same applies to the tectonic plates and *their* global possibilities and limits. Likewise can sea and land breezes be anything but local in their incidence? Or the Asiatic monsoon, well, "Asiatic" and therefore continental rather than global in scale? Drainage basins have their own scale hierarchy, the basins of tributaries nesting within that of the river to which they are tributaries.

Another way of putting this difference between human and physical geography is in terms of the relation to energy. The energy available to natural systems was for long a constant. It is now changing as a result of human intervention into the environment. Through their social development human beings have a unique capacity to increase the amount of energy that is available to them, whether through the use of fossil fuels or through harnessing the energy of natural systems: so geothermal power, wind power, solar power, and so on. The use of fossil fuels is clearly having an effect, through the warming of the atmosphere, on the energy that natural systems can use. It is suspected that one effect has been an increase in the intensity of storms. This may be related in turn to the warming of the oceans. Less equivocally, global warming seems to be having effects on tree growth. But the point remains: natural systems cannot act in a purposeful way so as to capture increased energy.[3] On the other hand, humans are certainly capable of doing so, and any increase in the energy available to climatic, hydrological, geomorphological or biotic systems is, apart from changes in solar radiation, ultimately due to that intervention.

Implications for the Knowledge Process[4]

The fact that people "act" while the objects of physical geography "behave" has major significance for the processes of acquiring knowledge about them. In talking about inquiry, the assumption has always to be that no-one can have direct access to the world, regardless of whether it is physical or social. Our knowledge is *always* mediated by concepts. To think otherwise is to commit the empiricist error. We have ideas about streams and what constitutes a stream just as we have ideas about spatial divisions of labor. We can never know for sure that our knowledge has captured the real world, though some of our knowledge can be put to practical test: try making water go uphill. On the other hand, our concepts change. Our understanding of what constitutes a stream, the nature of streams has been transformed; we now know that streams, contrary to what was once thought, increase in velocity downstream as a result of the declining significance of friction with respective beds and banks as the volume of water increases. We also know that, depending on the scale of resolution at which we measure it, its length varies: a finer scale of resolution takes in the meanders of the Mississippi while a coarser one will fail to. It took the idea of fractals, though, to underline that possibility.

Developing knowledge is a social process. As geographers we are inducted into – and disciplined into holding – a particular set of ideas in the world. Concepts are developed, disseminated, agreed on, though often with resistance. In this social process of conceptual and theoretical development we have to understand each other. Communication and therefore interpreting others is crucial, as though are the power relations within "disciplines": who gets to decide what is acceptable as human or physical geography. These claims apply to *both* human and physical geography and help us understand their respective histories.

However, there is also an important difference between them because of differences in their objects of knowledge. In human geography these objects are social in character: workers, owners, peasants and the like. Just as social scientists communicate with one another and develop a shared set of understandings about their objects of interest, so are those objects of interest developing their own set of meanings about themselves and those they interact with, including academic geographers. In other words: In human geography one not only needs to interpret the meanings of the concepts circulating among other geographers; one also has to bring those meanings into a relation with the meanings that our objects of knowledge assign to the world and to their interactions one with another. They can be two very different sets of meanings: something that has been taken advantage of, often very explicitly, as in the work on mental maps.[5] Accordingly human geographers have to engage in what is referred to as a double hermeneutic. The hermeneutic refers to the act of interpretation and human geographers are involved in two of these. Just why do people do the things they do? What are their reasons? We cannot stop at "causes" given by some social theory since these can be quite at odds with the meanings assigned to the world by the human practitioners themselves.

Physical geography is clearly different. Here, only a single hermeneutic applies. Physical geographers need to understand each other and arrive at some shared and accepted meanings. Their objects of knowledge, however – packets of air, streams, glaciers, terrestrial objects with different radiative properties – do what they do in a meaningless world. Theirs is not to interpret things and then act. Glaciers and rivers do not have reasons for what they do. Rather they behave. To use words commonly used in physical geography they are "forced" to do what they do as a result of "controls." Air ascending over a mountain range is forced to ascend in virtue of a geography of controlling conditions: high pressure on one side of the mountain range and low pressure on the other. An air mass does not hesitate when approaching the mountain range and say to itself: Do I really want to do this?

On the other hand, and regardless of their different subject matters, the social relations in which human and physical geographers are embedded are very similar indeed. There are the power relations within the subdisciplines that I referred to earlier. There is the same struggle around bodies of ideas and associated struggles for power among the representatives of those theories and practices: struggles for visibility within the discipline, around the biases of particular journals, sometimes leading to the foundation of new ones,[6] and for the honors that disciplines as social institutions distribute. There is also a set of wider social relations with respect to which any knowledge process has to be situated. Part of the "development" thrust in society as a whole is that knowledge should be useful outside the academy. There is massive pressure behind this attempt to channel intellectual activity in the academy as universities increasingly model themselves on businesses. Geography has been affected like other fields, and we all know the implications of that in terms of the sorts of direction in which that leads and the challenge of teaching students who are equally beholden to the same compulsions. Applied knowledge is usually contrasted with "pure" or knowledge for its own sake. What exactly does knowledge for its own sake mean? I would suggest that it means the pleasure of discovery about the world, the aesthetic stimulation of the pattern and order that research can uncover. This is development, but it is a different sort of development.

When we turn to how power is implicated in respective objects of analysis, there are important contrasts which flow from their differences. Power is central to the production of particular concrete human geographies. Human geographies are always for someone and are contested: think obvious cases like Brexit, residential exclusion, and sunbelts vs rustbelts. This is reflected in the discourse of human geography. There are sharp divergences of approach and, whether that is acknowledged or not, they are political to the core. The mainstream of professional human geography is complicit in what is happening. It is unaware of it because it is soaked in the ideological forms with which society presents itself and which are extremely difficult to avoid. Not the least of these is the idea that the social world comprises a set of independent, interacting forces: as in the classic regression model of dependent and independent variables, but the fundamental form of which has endured despite the relegation in status of the quantitative work. The obvious contrast here is with the totalizing – not deterministic – framework of Marxist geography which generates interpretations quite different from those of the mainstream, and even so-called critical geography (Cox 2016: 167–177). The result is that while it is easy to say that geographies are expressions of power relations, how those power relations are understood can vary very, very considerably: critical human geography is strong on distributional struggle, as in most of the writing on gentrification, while Marxist geography sees those struggles more as artefacts of ones that go on in the sphere of production and which condition the distribution of income.

These issues are far less clear in physical geography. In a famous essay in 1973 (127–128) Harvey drew contrasts between the social and physical sciences which have some bearing. Contestation around frameworks of analysis was clear in the social sciences to a degree that was not the case in the physical sciences. To the suggestion that this was because the social sciences were prescientific and had yet to find an objective basis for their study, Harvey suggested that, on the contrary, the physical sciences were pre-political. And the reason for this was that their control by a particular interest group had never been challenged. This was the reason that the pattern of revolutionary thinking followed by counterrevolution in the social sciences was not apparent in their physical counterparts. Of course, there were, and continue to be, revolutions in the physical sciences, but they pose no threat to the existing social order and so generate no counter-action.

One can readily see the possibilities here, and they have been grasped in part by the critical physical geography that we reviewed in the first chapter. I would suggest that the idea of knowledge for itself, at least hitherto, has as a necessary condition, a class relation. When one examines the great scientists of the nineteenth century, like Humboldt, Wallace, or Darwin, it is remarkable how they were the beneficiaries of flows of revenue, largely from land and investments (Darwin benefited from his father's investments in canals and property) so that they could afford to do their research without worrying about a wage or salary. On the other hand, the great pioneer of ecology, Gilbert White was a Church of England parson. We should not be surprised that two of the earlier books on geomorphology – *The Scenery of England* and *The Scientific Study of Scenery* were authored by people of independent means: Sir John Lubbock in the first instance and J. E. Marr in the second. Lubbock was the legatee of banking fortunes while J. E. Marr was a Fellow of St Johns, Oxford, and, one has to infer, supported by the college's very substantial property holdings. This would be an enduring pattern. Those secure in university employment, at least prior to the corporatization of universities, often wrote for wider publics with no thought of some materially purposeful application. The New Naturalist series put out by Collins in the 50s with a view to a lay readership had a number of contributions from physical geographers. This sort of venture was then complemented by a reading public, particularly of those who had been to university, that increasingly had the leisure for exploration of the outdoors, via rambling societies, holidays and days out in the country: so the geography of a place as something to be observed, as something separate, an object of consumption.

Concepts of Space in Human and Physical Geography

Spatial Concepts of a More Concrete Sort

One expression of the difference in the materials and processes studied is that while both physical and human geography share spatial concepts at a high level of abstraction, at more concrete levels there are ones that they do not. The concrete overlap is therefore a limited one. One talks about locating factories or highways but not about locating different sorts of landform or storm. One can certainly refer to the location of a range of mountains but the active sense of "locating" seems inappropriate. This is because of the way the verb has acquired the sense of some choice, which can only, with some very limited qualifications about animal behavior, be human. Likewise, when discussing the processes of interest of the physical geographer – processes of erosion, transportation, and deposition or of the movement of air masses – no-one talks in ways even remotely analogous to the ideas of segregation, exclusion and inclusion that human geographers use when discussing territory. Furthermore, in their concrete application, what seem to be similar concepts turn out to be quite limited in their similarity. Physical geographers talk about networks, but these are the fluvial networks of drainage basins where tributary streams flow into larger tributaries which then join to form an even larger one before debouching into the ocean. Morphologically there is nothing like this in human geography, unless it is the supply chains of firms and concepts of first and second tier firms.

In terms of spatial concepts human and physical geography vary in another regard. Space has three dimensions. For the most part, human geography confines itself to just two of them: the horizontal bits as in locating events according to a system of rectangular coordinates. There are some exceptions as with practices of transhumance or the upward and downward extension of the built environment of cities. Linda McDowell (1997) has written about the way in which corporate status is symbolized by how many floors up you are in the tower. But they are indeed exceptions.[7] This is not the case in physical geography. Rather the vertical component is also, and of necessity, a very significant one, omnipresent in its research.[8] Land masses are "uplifted" or they "subside" as in the great rift valleys of the world. The water of the oceans is layered in terms of its temperature, and the upwelling of cold water from below or the sinking of cold water can have important implications for the circulation of the atmosphere. And the latter, of course, proceeds at and between numerous different levels all the way from the storms encountered at the earth's surface to the jet stream which circles the globe in both hemispheres. Sea levels rise and fall and this has effects on stream erosion; as they fall, so erosion receives an added impetus. Increasing sea levels, on the other hand, swing the balance in the direction of deposition. Finally, we should note how the hydrological cycle through which one can connect both geomorphology and climatology is impossible to think of except in three dimensions.

This has implications for our understanding of scale. As we go from micro to macro, there are changes in both horizontal and vertical scale, as well as duration (Table 2.1).[9] Something like a depression not only covers a larger area than the typical land or sea breeze but also has greater vertical penetration and duration in time. The global circulation of the atmosphere is large scale both horizontally and vertically – think jet streams, for example – and changes according to an annual cycle.

There is something very similar in geomorphology. Summerfield (2005: 403) has suggested the emergence of two disconnected geomorphologies: "– one concerned with small-scale surface processes focused on short time scales and environmental applications, and the other a reinvigorated regional to continental-scale geomorphology integrating tectonics and surface processes to explain modes of landscape evolution." An emergent continental- even global-scale geomorphology is large scale in both horizontal and vertical dimensions and over time, in contrast to the geomorphology of an earlier generation, including Strahler, Chorley, and Schumm where the focus was the relatively small-scale study of fluvial landscapes in abstraction from earth movements and from the longer term[10] (see Table 2.2). Figure 2.1 illustrates the juxtaposition of the micro- and the meso-scales in landscape formation.

Table 2.1 Scale and climatic processes

Process	*Examples*
Micro-climatic	Land and sea breezes; fog formation; urban heat islands
Meso-climatic	Depressions; monsoons; orographic rainfall
Macro-climatic	Global circulation of the atmosphere

Table 2.2 Scale and geomorphic processes

Process	*Examples*
Micro-geomorphic	Formation of alluvial fans; lake infill; cirque formation; hydraulic geometry
Meso-geomorphic	Formation of fluvial/glaciated/desert landscapes
Macro-geomorphic	Tectonic movements

Figure 2.1 Micro-geomorphic and meso-geomorphic processes in juxtaposition. The landscape is one of the glaciated mountains of the Sierra Nevada of California. In the center of the image an alluvial fan has formed; one aspect of the infilling of the lake.

Source: Personal photo.

Although one does not typically think in these ways in human geography, an analogy seems worth exploring. Is it that the larger scales at which human activity organizes itself are more enduring over time? Capitalism as a system that is organized on a global scale has been around for about 350–400 years, but within that span of time, its uneven development has assumed variable concrete expression: not least, the emergence of newly developing countries and the decline of early comers. The same applies to countries: fairly durable frames of political economy, even while their position in flows of value on a global scale shift; but within them, the rise and fall of particular regions; and within regions as relatively more durable features, changes in urban hierarchies.

The difference in how space is conceived is also apparent in the contrasting graphical means that human and physical geography draw on in order to communicate their understandings and findings. Both use maps. But in physical geography there is a typical recourse to cross sections or 3-D

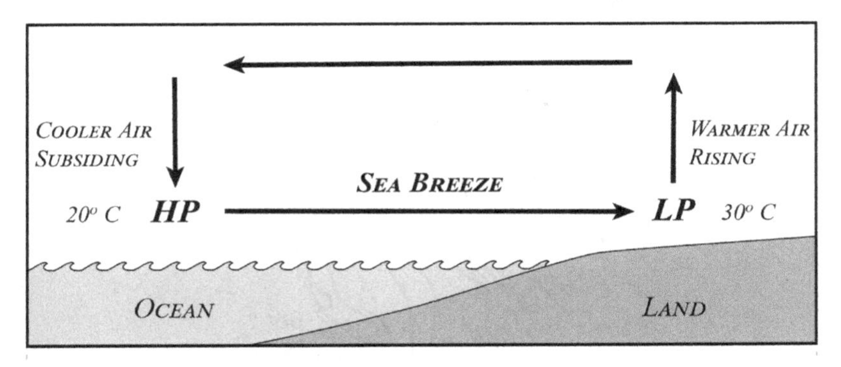

Figure 2.2 The sea breeze mechanism. During the summer the land heats up more quickly than the ocean. The warm air rises and is replaced by denser, cooler air from the ocean. See Figure 2.3 for the effects of this on the geography of temperature.

Source: Diagram by Jim DeGrand

diagrams. Three-dimensional diagrams abound in textbooks on physical geography but, and quite reasonably, they are almost entirely absent from human geography. Consider also the cross sections that are drawn on: cross sections of storms to illustrate their three-dimensional structure; cross sections of the atmosphere to show the way in which air returns whence it came but at a higher altitude (Figure 2.2); or cross sections of topography to bring out the relation between topographic form and the underlying geology – the alternation of scarps and dip slopes that one finds in parts of lowland Britain or in the Paris basin (Figure 2.4), or anticlinal formations where the top has been eroded away to reveal a series of scarps and dip slopes facing each other. A classic case is Southeastern England. The North Downs are of chalk and their scarp slope faces that of the South Downs – also of chalk some 20 or 30 miles to the south. In between are other reciprocally facing scarps of sandstone separated from the chalk by clay deposits (Figure 2.5).

Space and Disciplinary Identity

All this said, there is yet another fascinating contrast. A distinctive feature of practice in human geography is its concern with space, almost to the point of obsession. This was particularly so during the spatial-quantitative revolution of the 60s, but space endures as a criterion of what is geographic and there are continuing and energetic efforts to spatialize; testimony to the way in which the social theory that human geography draws on, has

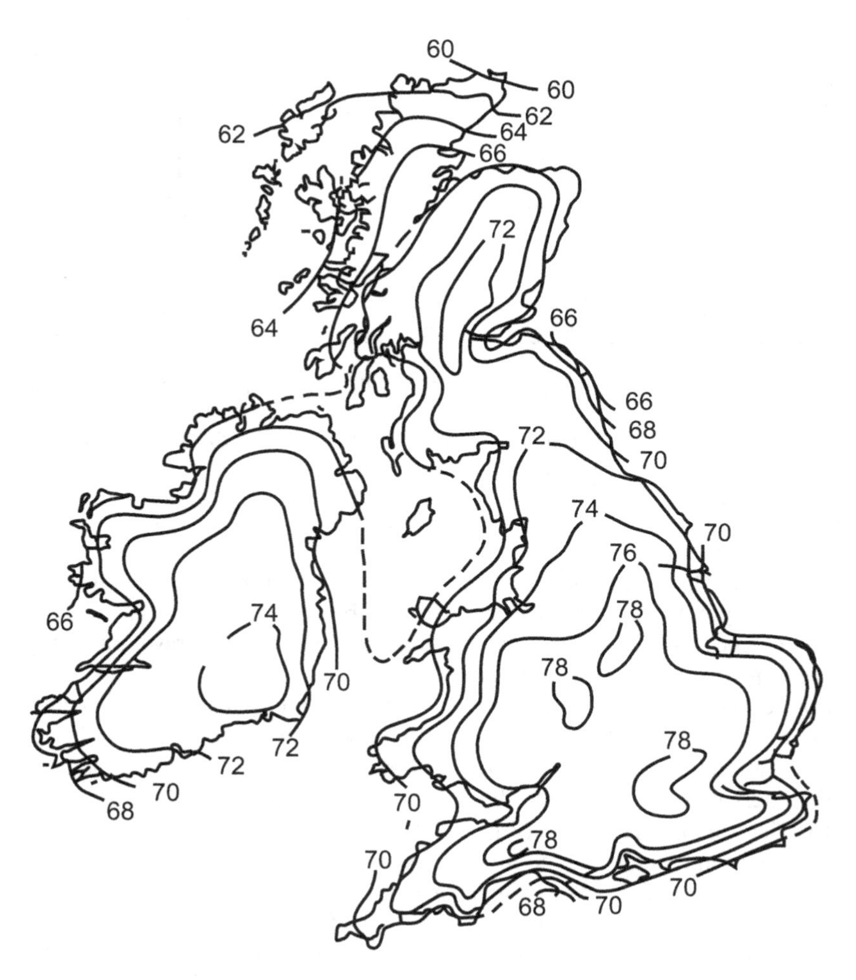

Figure 2.3 Effects of sea breezes on the temperature distribution: markedly lower daytime maxima along the coasts. July 1934; anticyclonic conditions.

Source: Manley 1952: Figure 52, p. 160.

remained quite aspatial. Crucially, though, there is no such parallel interest in physical geography, even while it is intensely spatial in its thinking and in its practice. So while human geographers have analyzed space and differentiated it into absolute, relative and relational forms (Harvey 2006), for physical geography it is as if it had not happened, even while these ideas

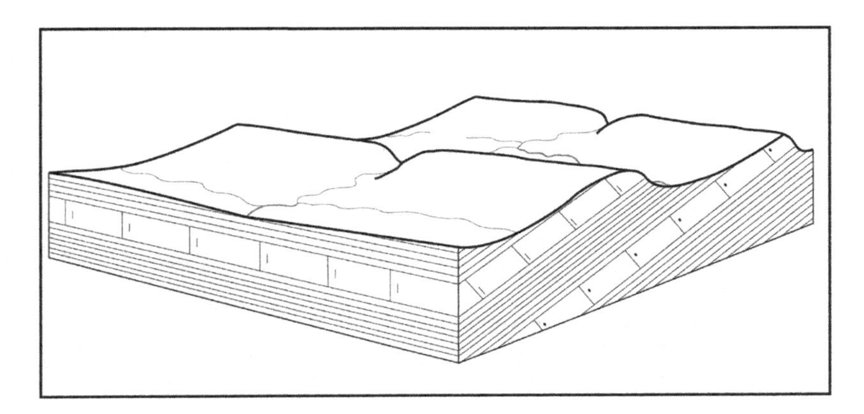

Figure 2.4 Scarp and dip topography.

Source: Diagram by Jim DeGrand, after A E Trueman 1949: Figure 7, p. 31.

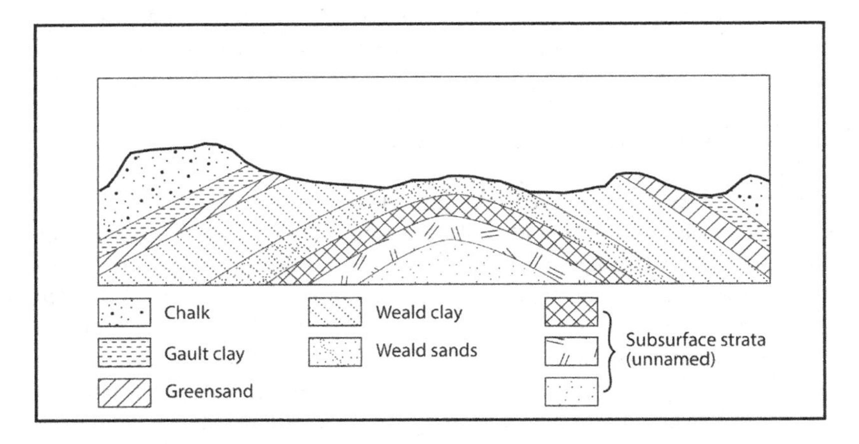

Figure 2.5 An anticline eroded from the top. The alternation of more resistant and less resistant rocks has produced a series of scarps and dips separated by an area of clay.

Source: Diagram by Jim DeGrand, after A E Trueman 1949: Figure 27, p. 99.

have as much applicability to physical geography as to its human counterpart and are (implicitly) drawn on. Landscape has a clear geometry and climates exhibit a spatial order. Space is an essential aspect of process in both human and physical geography. People move around leaving an imprint, as do air masses, if a more fleeting one. Configuration counts as an essential condition for understanding process in both subfields. If space enters

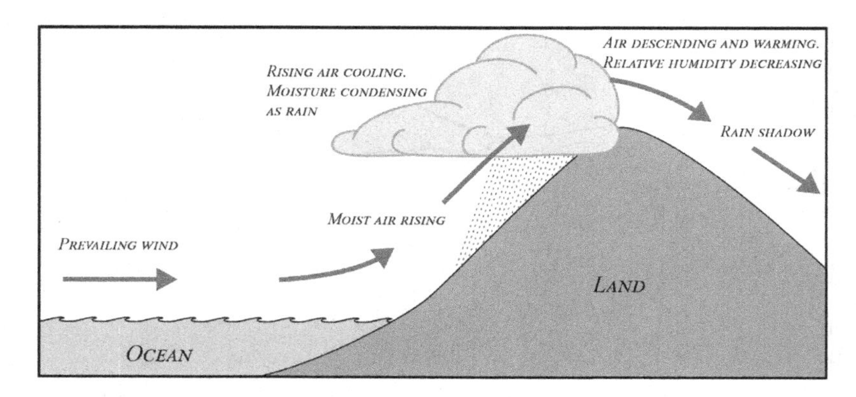

Figure 2.6 The rain shadow effect in cross section.

Source: Diagram by Jim DeGrand

explicitly into the physical geographer's thought process it is always in a highly concrete way, as in the effect of the spatial arrangement of rocks of differing resistance (Figure 2.5); or in rain shadow effects (Figures 2.6 and 2.7). The same applies to their approach to scale: none of the obsessing and contestation that occurs in human geography, therefore.

The difference is that in physical geography the relation to space is a constant one: air moves from areas of high to low pressure; streams flow downhill; land surfaces heat up and cool down more rapidly than water and this has effects on atmospheric circulation, as in daily sea and land breezes; sediment borne by incoming streams is deposited in lakes; and so on. In human geography, because of the distinctive qualities of people and their ability to make their own worlds through cooperation with one another, it seems otherwise. From this standpoint it makes perfect sense for human geographers to talk about locating things, while for physical geographers that is not the case, except for identifying the locations of tropical storms, rivers and the like, as remarked earlier. On closer examination, though, space is just as much a necessary aspect of social as it is of physical processes. It is just that social processes change and in consequence, so do locational propensities. Capitalism has its own spatial logics and these are transformed as, for example, technology changes, so that new spatial orders succeed one another. Human geographers like to fit gravity models to movement data, but over long historical periods the so-called distance exponent has tended to decline. People move over longer distances in virtue of social and technological changes. Economic geography is full of dramatic shifts as a result of technological change – as in the changes from water power to steam power to electricity – or through

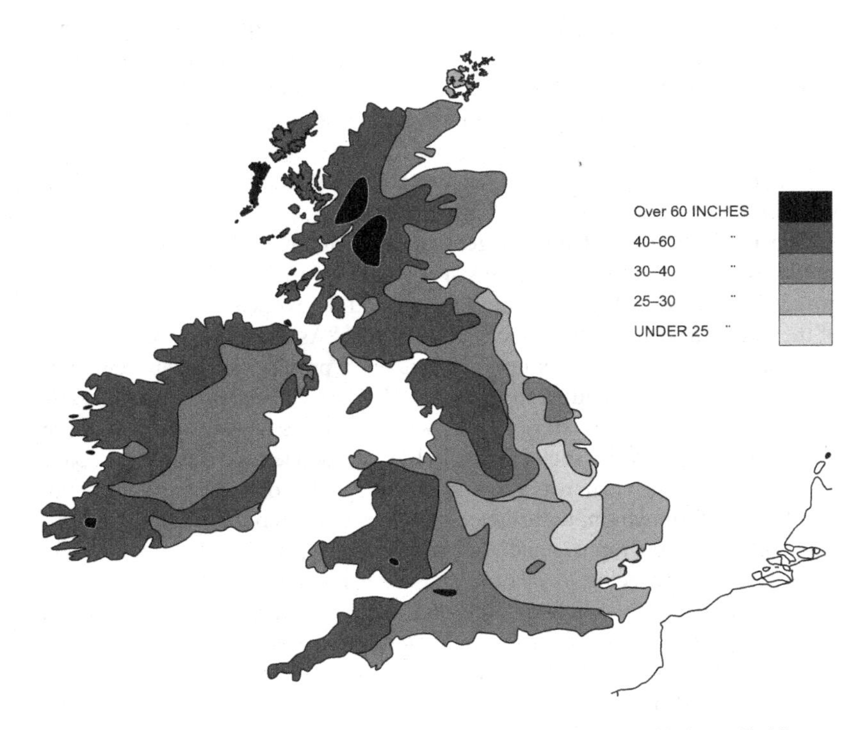

Figure 2.7 The rain shadow effect in England and Scotland. The higher relief lies on the western side of the island. The prevailing winds, heavily moisture-laden in these latitudes, are coming from the west. Central and eastern parts of the island are therefore in the rain shadow. After Mackinder 1907: Figure 83, p. 164.

innovations in the organization of production, as in "just-in-time." There is no equivalent in physical geography; no self-intervention on the part of elements of physical systems to change relations to space.

In physical geography the constant nature of the relation between space and process means that there can be no useful separation of the two. Attempts to separate them can produce questionable results. In a well-known and in many ways, useful paper on the development of the theory of continental drift and its transformation into that of tectonic plates, Jerome Dobson (1992) distinguished between what he called "spatial logic" and "process logic." In the original formulation of the theory of continental drift, and according to Dobson, Wegener relied on the former: the way in which some of the continents seemed to "fit" into one another – most notably the east coast of Latin America into the Gulf of West Africa; and the suggestion of continuities in the geology and fossil evidence between (once adjoining) continents. But according to Dobson, what was needed to clinch the argument was some evidence of a process that could allow continents to split off from one another. This would be at the heart of the theory of tectonic plates. And indeed it is a process, but it is a *spatial* process: the divergence of flows of magma under the ocean floor on which the plates essentially float. In physical geography, spatial logics are process logics and vice versa.

One might argue that human geography is different. But it needs to be emphasized that the social is always spatial and vice versa. Soja (1985) transposed Giddens' (1984) structuration theory and its explication of the relation of the individual to society to society and space: "As a social product, spatiality is simultaneously the medium and outcome, presupposition and embodiment, of social action and relationship" (p. 98). Any social action necessarily entails space as a condition: not least, social relations are always over space and occupy space; while social action in turn reproduces and transforms space relations. What this meant was a mutual presupposition of space and society but which, in its formulation, unfortunately managed to keep them apart. In short, unlike physical geographers and their physical processes, human geographers have had difficulty seeing social process as *necessarily* spatial: but then physical geographers have never had to confront a world in which its physico-(spatial) processes have continually changed in their form and within our lifetimes.

Space enters in not just through process: the flow of water over the landscape, the movement of air-masses, the spread of seeds via wind and the digestive systems of birds, the movement of people and commodities. The configuration of things in the world also makes a difference, but again, given the nature of the objects of study, this is something that varies between human and physical geography. In physical geography it occurs, for the most part, naturally: no sea breezes except where the land meets the sea; no rain shadows except where prevailing, moisture-laden winds cross a range of mountains; and no scarp and dip topography (see Figure 2.4) unless the strata are configured in a certain way; no monsoons of their current intensity with a different arrangement of the continents or the breakup of Asia in some way. These effects are entirely predictable and unchanging.

In human geography, it is otherwise. It is known that the configuration of places affects the parameters of movement between them. This is something that Peter Gould (1975) dubbed, after the person who originally identified it, the Curry effect: the distance exponent of the gravity model is sensitive to the geographic arrangement of origins and destinations. But it is also the case that those coefficients can change over time, perhaps as a result of changing modes of transportation. In some instances configuration is the focus of deliberate human intervention, always socially informed, in order to achieve some useful effect: the creation of new towns in order to reduce pressure on metropolitan housing and labor markets; or the just-in-time arrangements of supply chains in industry to increase the velocity of capital and hence the profit it can turn in a given period of time (Sayer 1986).

There is also human intervention into natural systems in order to create new configurations that will work to human purpose. They can have unintended effects: the concentration of buildings and concrete with a high ability to store the heat of the sun's radiation and then radiate it back creates the well-known phenomenon of urban heat islands and, as remarked earlier, urban sprawl undoubtedly aggravates global warming (Sayre 2010). But there are clear limits to this. The arrangement of the continents is crucial for climate geography and according to some, for geopolitics, but there is no geoengineering yet available through which one might redirect the movement of the tectonic plates. California may ultimately break away to the pleasure of some and to the disappointment of others, but it is not something that human endeavor can promote.

One way of understanding this difference is in terms of a contrast between external and internal necessity.[11] Human geography, at least under capitalism, is the sphere of external necessity, and physical geography, internal. The social relations of human geography are just as necessarily spatial as are the purely physical ones of physical geography. It is just that in human geography, in virtue of social change, the relationship is a shifting one and is experienced as something external: something that has to be sought out in order to take advantage of, say, some technological change. The shift in the form of the factory from multiple floors to one of a single floor to accommodate the assembly line, could not be accommodated in central cities, pushing them out to more peripheral sites: a necessary relation but also an external one, in that it had to be sought for and discovered. In contrast, in physical geography, there is no searching out for a spatial fix; the spatial fix is inseparable from processes that are unchanging.

There is a sense in which the way in which physical geographers have understood space has been, for the most part, in advance of human geographers. Harvey differentiated between three distinct concepts of space: absolute; relative; and relational. In its absolute sense, space is mere container or pigeon-hole for organizing information about the world: so, the Midwest, the Carpathians, the Sahara Desert. In human geography, this tended to dominate for at least the first half of the twentieth century, and expressed in a highly particular version of regional geography. Relative concepts of space would blossom with the spatial-quantitative revolution. The location of a city or area, relative to other cities or areas, became central to understanding their occupational composition or movements of people, commodities, and the like between them. In this conception, the quantifiably calculable attributes of places are contingent on their locations relative to others: change the locations and their attributes change. The discovery of relational space has been more recent. According to this, places internalize aspects of other places: their relation is a necessary aspect of what they are. London as a capital city is inconceivable outside of a country to govern, to pay taxes to keep it going. In spatial divisions of labor, the relations between places occupying different positions in it are necessary ones: no headquarters locations without production somewhere.

There is no hierarchical ordering here. The concept of space drawn on depends on one's purpose. For a travel agent, informing clients about the comparative advantages of different destinations in terms of climate, cost, etc., space as absolute is perfectly adequate, just as a real estate agent is likely to see geography through the prism of relative space: where a property is located relative to places of work, of recreation, shopping, school districts and the like. For analyses of a more causal sort, relational space is indispensable: how to explain places in terms of their substantive connections with other places. And taking the global, how they are necessarily capitalist sorts of places; which in turn helps us understand real estate agents and *their* particular geographic viewpoints.

These different conceptions also apply to physical geography (Table 2.3) but, again, physical geographers are blissfully unaware of the distinctions being drawn. Their history has also varied. In physical geography there was an embrace of relational space as a central concept, albeit implicit, long before it was recognized – more or less – in human geography. The fact that depositional landscapes depend on and reflect erosion elsewhere has long been taken for granted and often of crucial significance in reconstructing past landscapes.[12] Likewise from the air mass revolution on in climatology, the way in which regional climates were molded by air masses with diverse sources and trajectories (see Figure 2.8) became conventional knowledge.

Table 2.3 The different concepts of space as they apply to human and to physical geography

	Human Geography	*Physical Geography*
ABSOLUTE	Classical regional geography: France; East Anglia; the Ozarks; Hackney.	The Scottish Highlands; Patagonia; the Amazon basin; the Caribbean.
RELATIVE	central place theory; center and periphery; the gravity model.	Köppen's hypothetical continent; landscape geometry; rain shadows; the Wallace Line; continentality.
RELATIONAL	spatial divisions of labor; the hybridity of places; colonizers and colonized.	glacial erratics; air mass climatology; climatic geomorphology; depositional landscapes; ocean currents and regional climates.

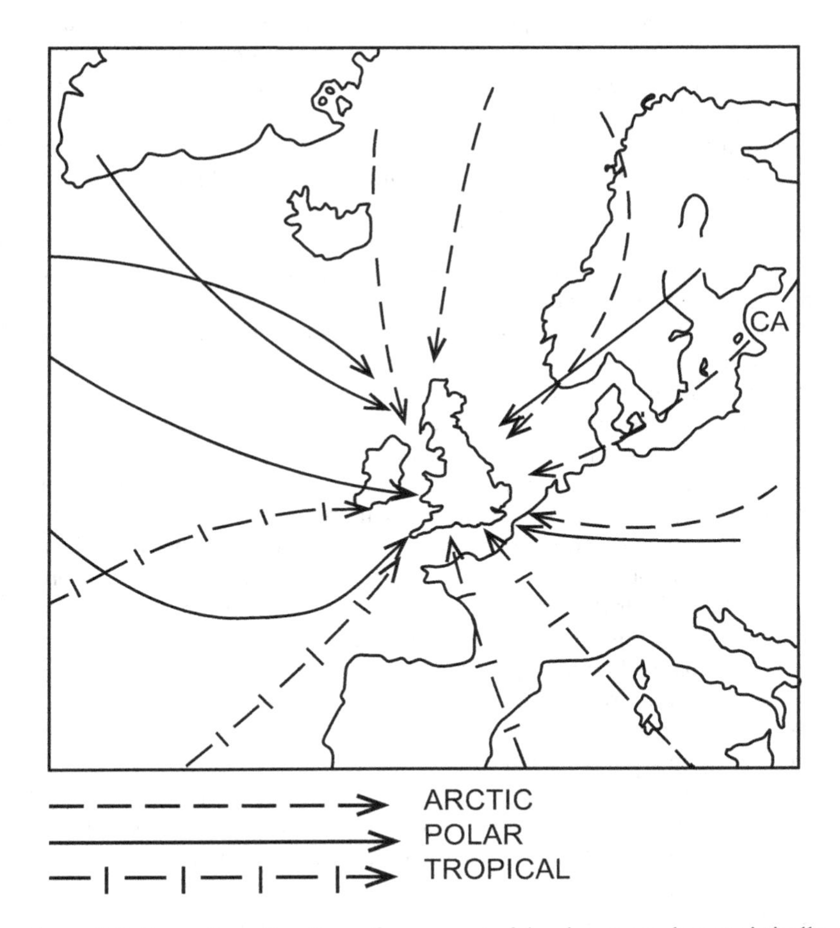

Figure 2.8 Generalized directions of movement of the air masses characteristically reaching Britain. After Manley 1952: Figure 15, p. 71.

Unity and Separation

Human geography differs from physical geography in another sense. The social process can be seen as a unity. This is a contested view but there is a case for it. In physical geography, it is clear that while in every case it is a matter of physical processes, these differ quite markedly between climatology, geomorphology, etc. There is no integrated physical process like the social process at the heart of human geography, that is even conceivable.

To say that human geography can be seen as a unity needs some explanation. Of course, as in physical geography, there are processes which can be regarded as contingent in their relationship with one another: the state has a logic different from the division of labor. But in the social process one can argue that there is a particular relationship which organizes and reworks these other logics to make them compatible with, and facilitative of, its own. This is the Marxist argument. Beliefs, the division of labor, institutions, including the state, the relation to nature, spatial arrangement, class relations, all have to be consistent with the demands of material production and reproduction, and these are always socially mediated: a function of what Marx called "production relations." This does not mean that coherence is a settled feature of social life. Rather, as the different moments of the social process pursue their own logics, so they come into contradiction with that focused on the production and reproduction of material life and coherence has to be reestablished once again.

In physical geography, there is nothing remotely similar. The different processes, climatic, geomorphological, biotic, hydrological, remain independent of one another. There are certainly relations between them. Landforms cannot be understood except in terms of climate, past or present. Continental drift changes the climate geography of the world. These, though, are relations of a contingent, non-necessary sort. The implications of climate for landforms depends, *inter alia*, on the underlying geology, on the preexisting topography and on crustal movement. Continental drift does indeed change climate geography, but only within limits posed by the earth's rotation around the sun and the annual march of the seasons.

There is also the fact that the time scales over which the processes act are quite different. For the larger part, physical landscapes are relict; change can be observable, providing clues as to how those landscapes evolve, but millennia must elapse before all traces of, for example, glaciation or uplift are eliminated. Climate is something else, responding rapidly to changes in underlying conditions, even while those conditions may change very slowly, though not necessarily, as in the case of urban climates. The same goes for changes in biogeography.

In human geography it is otherwise. The relations between the standard systematic divisions of economic geography, cultural geography, can appear quite arbitrary in a way which does not apply to the systematic fields of physical geography. Accordingly one can argue that the economic is always cultural/political/spatial/ecological/institutional, etc., and has to be theorized in terms of these other aspects of itself, and vice versa. Not all human geographers would agree with this and would defend a more pluralistic view of the world. But in the case of physical geography one cannot even begin to argue for, let alone defend, the existence of a master process that organizes and subordinates the others to its purpose. Accordingly, one can theorize processes in geomorphology without being a climatologist. All one needs are certain expectations regarding climatic baselines or trends, but not why those baselines or trends exist. So even though Malanson et al. (2014) talk about an "integrated physical geography" this is all it amounts to: a set of contingent relations where landforms provide a condition for the realization of climatic processes; and where climatic change is one condition among many to be taken into account when interpreting landscape evolution.

Summary Comments

There are radical differences in the nature of the objects constitutive of those geographies that are in question in human and physical geography (Table 2.4). These helps shed light on some curious differences in debates: human geographers have been extraordinarily interested in the difference that space makes and have written extensively on the matter. There is virtually no interest in the topic in physical geography. In physical geography the necessary connection between space and process, how process is always spatial and the relation is a constant one – air always moves from high pressure to low pressure – can be taken for granted. In human geography, there is a similar relation between social process and space relations. The problem is that as social processes change, so does the relation to space, which can suggest, and quite misleadingly, something contingent rather than necessary.

It is not, though, simply a matter of fundamentally different objects of study that keeps the two subdisciplines apart. Their conception of space differs in its emphasis. In physical geography, the vertical dimension of space can be seen as no more or less important than the horizontal ones. In

Table 2.4 Human and physical geography: some essential differences

	Human Geography	*Physical Geography*
OBJECTS	Social	Natural
METHOD	Double hermeneutic	Single hermeneutic
PROCESS TEMPORALITY	Developmental	Directionless; at best, cyclical
DIMENSIONS OF SPACE	Largely horizontal	Both horizontal and vertical
SPACE AS A NECESSARY RELATION	External	Internal
FUNDAMENTAL PROCESSES	Unified	Separated

human geography, and with a few exceptions, the two dimensions rule and verticality can be excluded.

So even while at a high level of abstraction we are talking about fields that focus on spatial relations, sharing concepts like circulation, diffusion, directionality, fixity, area, the more concrete forms of those relations vary considerably: continentality, watersheds, orographic rainfall, the potential energy of elevation as compared with, say, center-periphery effects, spatial concentration, and segregation. Bringing the two branches of the field closer together has to be in terms of shared space relations. The question is: How to do it in a mutually illuminating manner? It is to that question that we turn in the final chapter.

Notes

1 I am using the Marxist understanding of "development" here: that through the ability to produce on an increasing scale, people develop their understandings of the world, the institutional forms through which they produce, and in consequence their needs and their ability to satisfy those needs. The Introduction to the *Grundrisse* is particularly good on this; likewise the *German Ideology*.

2 Compare the famous statement from the opening paragraphs of Chapter 7 of Marx's *Capital Volume 1*: "Labour is, in the first place, a process in which both man and Nature participate, and in which man of his own accord starts, regulates, and controls the material re-actions between himself and Nature. He opposes himself to Nature as one of her own forces, setting in motion arms and legs, head and hands, the natural forces of his body, in order to appropriate Nature's productions in a form adapted to his own wants. By thus acting on the external world and changing it, he at the same time changes his own nature. He develops his slumbering powers and compels them to act in obedience to his sway."

3 The energy available to natural systems may change. With denudation and the reduction of the land surface, the potential energy of elevation is reduced; on the other hand, the uplift of the land, either through tectonic action or through the melting of ice sheets, can occur, with a resultant increase in the potential energy of elevation. And volcanic eruptions can interfere with the amount of solar radiation reaching the earth's surface and hence reduce the energy available for atmospheric systems. But in no instance is this due to any purposeful action.

4 On the knowledge process, see Sayer's highly illuminating discussion in Chapter One of his book *Method in Social Science*.

5 An interesting case, since if one reads these studies from the late 60s to early 70s, the underlying assumption is of a benchmark "map" that is objective in nature and not something that is conceptually mediated.

6 *Geographical Analysis* was a classic case; born of the frustrations of quantitative geographers trying to get their work published in the established journals.

7 A miniseries expressed a particular socio-spatial arrangement of a vertical sort through its title *Upstairs and Downstairs*. However, when I went to South Africa, I found that this arrangement was reversed and the (African) servants

lived on the flat roof at least in the case of Johannesburg hi-rise apartment buildings,: what were called "locations in the sky." According to Olivia Manning's *Balkan Trilogy*, this was also the arrangement in prewar Bucharest. Otherwise put: human geography is complicated by cultural practices in a way that physical geography is not.

8 Goudie and Viles (2010) express this in a particularly vivid way in their claim that "the essence of geomorphology can be interpreted to be the long-term interplay between climate and tectonics" (p. 8.); that is, by processes powered by energy from outside the earth and processes powered by energy from within the earth.

9 See Barry 1970: 62.

10 Compare Schumm and Lichty: "We believe that distinctions between cause and effect in the molding of landforms depend on the span of time involved and on the size of the geomorphic system under consideration. Indeed, as the dimensions of time and space change, cause – effect relationships may be obscured or even reversed, and the system itself may be described differently" (1965: 110). See also Martin and Church 2004.

11 The distinction is that of Marx: "The more deeply we go back into history, the more does the individual, and hence also the producing individual, appear as dependent, as belonging to a greater whole: in a still quite natural way in the family and in the family expanded into the clan; then later in the various forms of communal society arising out of the antitheses and fusions of the clan. Only in the eighteenth century, in 'civil society', do the various forms of social connectedness confront the individual as a mere means towards his private purposes, as external necessity" (Marx 1857–58: 84).

12 As in drawing conclusions about particular cases of river capture.

3 How Configuration Matters

> "Geography lies in the way that apparently disparate phenomena might be connected or contextualized. Geography might (if in analogy only) therefore be an 'emergent property' that can be neither predicted nor explained from its initial conditions, or from knowledge of its component parts. It is encountered rarely and may be short-lived."
>
> (Clifford 2001: 389)

Introduction

The most comprehensive attempts thus far at integrating physical and human geography are, as I suggested in Chapter 1, those of Bunge and Massey, even while they came from very different directions. But both abstract space relations without reference to other aspects of their objects of interest, and quite justifiably. The fundamental difficulty is ontological. The two subfields deal with very different sorts of object which makes developing a theory that embraces both – that is, the physico-spatial objects of physical geography and the socio-spatial objects of human geography – thoroughly problematic. These problems are graphically foregrounded by the fact that differentiation and spatial relations appear as necessary aspects of the relations with which physical geography deals; but that is not how it appears in human geography, or, as most famously and controversially argued by Sayer (1985), how it actually is.

The best that one can hope for, for the most part, is at an epistemological level and the exchange of models of understanding, which is the case in both Bunge and Massey. The one possible exception to this claim would be work focusing on human intervention into physical systems that we reviewed in Chapter 1 under the heading of "Man's Role in Changing the Face of the Earth," but so far these studies have failed to provide much in the way of an understanding of the part played by the socio-spatial; in other words, just why have people intervened in the way that they have? The indivisibility of

DOI: 10.4324/9781003362708-3

geography has its limits, therefore. Nevertheless, geography does illuminate the world that we live in, both human and natural. It is also a response to a deeply felt desire to understand the world around us, finding expression in, among others, an interest in landscape in all its facets, including, as Gordon Manley (1952) so evocatively noted, skyscapes.

What therefore might a shared framework that highlights the possibilities of a sharing of modes of understanding look like? Massey's arguments regarding human and physical geography help, but the sort of relationality she describes needs to be reconsidered in favor of Harvey's more dialectical approach: one where a world of "permanences," physical and social, is constituted by flows. It is also the case that she is too focused on complexity and that would exclude large areas of both physical and human geography.

To explore the possibilities further, consider at the start the traditional objects of interest in geography: cities, particular climates, cirques, regions, or even ensembles of objects like central place patterns, basin-and-range topography, or alternating land and sea breezes. They can all be understood as configurations in time-space, as "permanences," constituted by flows and relations: Cronon's (1991) Chicago as nature's metropolis; climates as functions of the distribution of land and water and the general circulation of the atmosphere; and landforms as conditional upon climate and tectonics as well as the mechanics of erosion and deposition – in short, how objects internalize a geography of flows and relations without which they would not exist.

To see them in this way is to emphasize their contingent nature, but also in some, though not all, cases, their complexity as particular juxtapositions generate nonlinearities in development and emergent forms. To understand why configurations have the effects that they do, we clearly need more general theory of a spatial sort: knowledge of the circulation of the atmosphere, the mechanics of soil creep, the rules governing the circulation of commodities, the hydrological cycle, the carbon cycle, all of which abstract from the particularities of configurations but which must always return to them for confirmation and revision. On the other hand, those abstractions also depend for their validity on more global sorts of configurations or conditions. The circulation of the atmosphere depends on latitudinal variations in net radiation but also on the tilted axis of the earth's rotation and the direction, west to east, on which the earth rotates. The circulation of commodities depends on the presence of capitalist relations of production. Potential elevation as a determinant of erosion depends, excluding lakes and inland seas, on sea level, and hence on changes in the earth's temperature.

A possible objection to this program is that it provides for no contributions of a more general or universal sort. This, though, is to overlook the way in which cases provide an important contribution to theory. This occurs in two ways: The first is through studies in particular areas: Christaller,

Southern Germany, and central place theory; Agassiz, North America, and the theory of continental glaciation; W M Davis, the Northeast of the United States, and fluvial landscapes; Bjerknes, Norway, and midlatitude storms; Lester King, South Africa, and the parallel retreat of slopes; Gottmann, the Northeast Corridor of the United States, and megalopolis. And secondly theory works, and complementary to the use of case studies, through analogy: examining particular configurations for similarities and differences with those located elsewhere or elsewhere at other times. What light does South African apartheid shed on Bolivia? What does the study of a particular climate in a particular place tell us about how that same climate works out somewhere else?[1]

Geography and Configuration[2]

In what is often regarded as a key article in the history of (human) geographic thought, Nigel Thrift (1983) drew attention to a particular distinction in how we approach the subject matter of geography. It turns out to be pregnant with implications for the current project. The contrast in question was between the compositional and the contextual or, otherwise expressed, the immanent and the configurational. The immanent refers to law-like statements about the world; the configurational, to how those law-like forces get expressed in particular contexts or configurations. Thus Hägerstrand in 1984:

> With his law of gravitation Newton could predict how an apple accelerates when it falls to the ground from its branch. Perhaps somebody has calculated the strength of apples seen as material. But none of these general pieces of knowledge are sufficient to tell if Newton's apple would get crushed or not after reaching the ground. In order to judge that part of the event, one needs to know if the ground is made up of, for example, grass or hard pavement. One must, in other words, look into the diorama which shows in what ways things are present with respect to each other just where events happen.
>
> (pp. 3–4)

Earlier, the geomorphologist Barbara Kennedy had made a similar point. In an article in 1979, with the intriguing title "A Naughty World," she drew on the work of the geologist George Simpson (1963) to critique work in physical geography that tried to reduce it to law-like statements; statements, in other words, that could provide predictability about the world. Her object of critique was the, quite transient, fascination with systems theory, at the time seen as something that could be practiced in both human and physical geography:

> The elements of the real world – both natural and man-made – which go to make up (the) environmental systems . . . are all products of their location in space and time; all have histories, yet all are also obedient to the fundamental laws of the universe. What has been very clear for a number of years now is that the *configurational elements* (my emphasis) which must be taken into account in attempts to explain the past, present and future behavior of the world in a geographic sense, are extremely difficult to handle in an explanatory framework which derives too narrowly from physics, where such concerns are generally defined as irrelevant.
>
> (1979: 552)[3]

It is important not to be distracted by the references to physics here. The applicability of the contrast is much wider than that: how the laws of chemistry, biology and society get expressed is equally subject to contextual considerations and this is fundamental to the practice of the spatial sciences. In those cases where practical applications are critical, it has been central to their self-identity: epidemiology is a classic case. In geography, the enthusiasms of the quantitative work of the 60s, in both human geography and geomorphology, tended to emphasize the immanent at the expense of the configurational. For a discipline historically dedicated to understanding differentiation across the earth's surface, this now seems a little odd: a brief moment of amnesia as the discipline struggled with its image by taking on a law-seeking, supposedly "scientific" position. Recognition of the significance of the configurational, on the other hand, underlines the unity of human and physical geography; and it can be expanded beyond the simple observations of Hägerstrand and Kennedy.

One of the first people to recognize this was Doreen Massey in her paper of 1999 (1999b) where she suggests that both human and physical geography should be redefined as "complex, historical sciences." The "history" part is evident from what has been said already: how processes work themselves out depends on configurations which are simultaneously spatial and historical: configurations have a history, so history has to matter in understanding geographic difference. How the laws of climatology work themselves out depends on configurations of land and sea, mountain and plain, the ebb and flow of solar radiation over the millennia, and these all have a history. Likewise, the configuration of transport networks and geographies of urbanization: whether highways and railroads were more or less centralized in their geometry has made a difference. If in doubt, compare France and Germany.

Massey, though, in an important intervention, goes beyond this rather simple notion of contingency to talk about configuration as having other effects.[4] Configuration does not simply affect how some process gets realized and the effect it has: whether or not Newton's apple lands on grass or hard pavement, that is. Through the interactions between its different parts configuration can give rise to entirely new structures of relations and processes: chance juxtapositions and processes of agglomeration in human geography, particular configurations of atmospheric pressure, ocean temperatures, and the generation of storms. History, or more accurately, geohistory, matters in an altogether more profound way than the simple contingent effects of configuration.

Difference in configuration is energizing. As Massey says (1999a: 274), ". . . time needs space to get itself going; time and space are born together, along with the relations that produce them both." Erosion requires the potential energy of elevation: differences in altitude; differences in receipts of solar radiation across the earth's surface are the condition for the different pressure belts which drive the circulation of the atmosphere; migration responds to differences in the demand for labor relative to its supply. And so on.

There are some limits to her understanding. She tends to neglect the significance of more universal conditions: the immanent as opposed to the configurational. If people are to act with respect to various juxtapositions of forces and conditions, they need some incentive in order to do so and motivations are typically conditioned by more universal sorts of social relation, like the family or markets. Likewise fluvial erosion depends on far more than local variations in elevation, not least the heat laws which determine whether moisture occurs in the form of water or ice. Air masses are affected by the different surfaces over which they move, but their general directions are determined by the laws of atmospheric circulation. Likewise, and in accordance with her emphasis on emergence, she wants to embrace ideas from complexity theory. It is not clear, though, that that is where all the emphasis should be placed. Contingency invariably enters in regardless of whether a particular spatial arrangement can be regarded as emergent or not. It imparts an historical element to geographic variation, therefore, regardless of questions of complexity, as we will now see.

Contingency in Geography

Human Geography

Over the past 50 years in human geography there has been a slow realization of the significance of configuration. Partly this has been in the form of a rediscovery of variation in the way spatial tendencies are realized: how fundamental parameters tend to vary and in intelligible ways. If one could give a date to this, it would be Rushton's paper of 1969 where he distinguishes between what he called "spatial behavior" from "behavior in space." Spatial behavior abstracts from the particularities of spatial arrangement, while behavior in space takes it into account – journey-to-shop patterns were necessarily conditioned by the distribution of retail outlets. This would then be confirmed by the critical work of Les Curry (1972) and then Ron Johnston (1975) on how the distance coefficient in the gravity model tended to vary as a function of the arrangement of points of origin and destination with respect to one another. That these could then be rationalized was shown by Fotheringham (1981) in a study of air traffic from different points of origin in the United States: marked boundary effects so that the origins on the margin had smaller distance coefficients in accord with the need, simply, to travel further to get to the rest of the United States. Gould (1975) would dub the effect of spatial arrangement on the distance coefficient "the Curry effect." He then showed regularities in its geography for the case of information flow much as Fotheringham (1981) did for airline traffic.

There was earlier work whose more general theoretical significance had gone unremarked. In his work on central place patterns, Brian Berry (1967) emphasized how spacing depended on population density: in North Dakota, one simply had to go further to get what was wanted than in Illinois or, better yet, Chicago. Peter Taylor (1971) then did some wonderfully innovative work looking at how the shapes of countries affected migration distances: so, and taking into account the distribution of population, the relative balance between long and short distance moves *had* to be different in a country as elongated in its shape as Italy than in a more compactly shaped country like France. But in neither case was the idea of contingency foregrounded. The lessons drawn were different. Berry wanted to show how the principles of central place theory held up regardless. Peter Taylor focused on shape, how it could be measured and why those measures might be useful in understanding the frequency distributions of movements.

The idea of contingency and its radical implications for spatial theory, would actually come from outside the spatial-quantitative work. This was through the contributions of Andrew Sayer (1992). Sayer tried to draw lessons from critical realism for human geography. A crucial distinction was between causal necessity on the one hand and how that causal necessity was realized in concrete terms; the latter was a contingent matter – possible but not necessary, which recalls Hägerstrand's story about Newton. This allowed some critical traction with respect to some of the conclusions geographers had come to: tendencies to read the world directly off from theory; or to move in the reverse direction and to theorize on the basis of cases that were more complex in their causal structures and owed much to contingent conditions. It also allowed some convergence with spatial-quantitative work and a realization there that correlations and regressions needed to take into account the role of the contingent (Jones and Hanham 1995).

However, critical realism was controversial in some quarters. Harvey (1987) objected and, as I have outlined elsewhere (Cox 2013), I now think that he was correct. From the standpoint of social theory, the distinction between the necessary and the contingent is too sharp: what is contingent can be drawn on in the pursuit of social necessities and as such, can become necessary in its turn. There is, though, still a case to be made for pure contingency in physical geography, as we will see. But in this regard, human geography is different. This is the reason why, as noted in Chapter 2, unity of process in physical geography has remained elusive.

Something else needs to be noted at this point: how contingency and therefore configuration have been incorporated into geographic research. They figure largely as ways of qualifying general laws. Contingent effects can then be drawn on as themselves objects of generalization as in Fotheringham's (1981) work on distance coefficients. In some of his work, though, Sayer, perhaps inadvertently, pointed in a different direction. This was in the context of a paper on regulation theory calling attention to the way in which Japan was quite different and suggesting that countries themselves, and drawing on the organic character of their societies, could be objects of analysis in their own right.[5] What he did not note was that behind Japan's distinctiveness lay particular configurations of forces and conditions that had then interacted to give rise to it as we currently know it: a set of mutually supporting and reinforcing conditions. This is a direction of analysis that can be generalized across geography as a whole.

Physical Geography

What has happened in physical geography has varied. Geomorphology has in its own way, perhaps in part as a result of its closer association with human geography, but also for historical reasons, moved to embrace the importance of the contingent and configurations of conditions that make a difference to how more general processes play out. Critical realism has even put in an appearance (Richards 1996;[6] Rhoads 1994; Slaymaker 2009). Climatology has been different, and one can speculate why that is. Unlike landforms, climates do not bear the imprint of their history, except over very short periods of time, though to be fair, contingency does figure: the arrangement of land and water on a global scale is immensely important for how the laws of atmospheric circulation play out in their effect on regional climate; the evolution of life on earth has also had major implications for the chemical constitution of the atmosphere.

To focus on work in geomorphology, according to Phillips, "Landscapes are indeed shaped and controlled by deterministic, global laws, but the operation of these laws in specific geographical and historical contexts means that landforms and landscapes are often circumstantial, contingent outcomes, not derivable from global laws alone" (Phillips 2007: 167). He has argued this at considerable length, pointing out the various contingencies in play: not least lithological, climatic, and tectonic (see Figures 3.1.1 and 3.1.2). This is a view shared with others, and in part as a more general reaction to a changing balance of more general viewpoints in geomorphology: from an emphasis on process studies to greater concentration on landscapes in their totality, where historical considerations have to play a larger part; but now in a context where techniques of dating facilitate their determination in a way previously impossible (Summerfield 2005); and also where there is, as a result of the past experience of what Strahler called, "a dynamic approach to geomorphology," a greatly enhanced awareness of how more general processes work. People now talk explicitly about spatial arrangement and its significance, albeit in the context of a recognition of time as well.[7] However, and again, the idea is being drawn on not to shed light on the totality of landscapes as Phillips indicated by his summary term "the perfect landscape," but to shed light on how general processes combine with specific juxtapositions of conditions and influences in time-space to affect the development of particular sorts of landforms: so, and for example, meander forms or alluvial fans, or

Figure 3.1a: Geology and landscape: Contrasts in Central Idaho. The two landscapes depicted are both glaciated and within ten miles of each other. One (3.1a) consists of volcanic rocks; the other, sedimentary rocks. Sawtooth Mountains, Idaho.

Source: Personal photo.

Figure 3.1b: Geology and landscape: Contrasts in Central Idaho. The two landscapes depicted are both glaciated and within ten miles of each other. One (3.1a) consists of volcanic rocks; the other, sedimentary rocks. White Cloud Mountains, Idaho.

Source: Personal photo.

indeed contrasting glaciated landscapes in Idaho illustrated in Figure 3.1.[8] On the other hand, what Summerfield said in 2005 has yet to be brought to fruition. He talked there about two geomorphologies: " one concerned with small-scale surface processes focused on short time scales and environmental applications, and the other a reinvigorated regional to continental-scale geomorphology integrating tectonics and surface processes to explain modes of landscape evolution?" (p. 403). One can fairly wonder where the hoped-for integration appears in the published research.

There are similar research patterns in climatology. The contingent effects of relief and of the arrangement of the landmasses and oceans has long been recognized and researched: how, that is, they make major differences to how the dynamics of the atmospheric circulation – predictable to a very considerable degree as in the recognition of the different pressure belts – affect climate. The idea of continentality and how it influences climatic variation is an old theme; likewise orographic rainfall as in research into the various conditions that can give rise to it. But again, as elsewhere in geography, it seems that contingency is recognized only to the degree that it deforms the expression of more general laws. The notion of configuration as underlying the formation of something regionally distinct gets short shrift.

To talk about "Mediterranean climates" is to engage in the crudest of generalizations. The Mediterranean climate of the Californian coast is different from that around the Mediterranean itself as any European visiting Southern California in the months of June and even July will quite painfully discover: foggy mornings deter lying on the beach. And the reason for the fog has to do with the fact that there is a cold ocean current offshore not replicated in the Mediterranean. And there is more. The flipside of this for the heliophile is the fact of the Santa Ana: again something not replicated around the Mediterranean basin. Rather the seasonal wind that blows in the northwestern part of the Mediterranean basin is the cold Mistral. Then throw in the periodic effects of El Nino on California and the differences become even more stark.[9]

Hare remarked thus on the distinctiveness of the climates of North America:

> The climates of North America are also unique in the high degree to which they depend upon facts of relief and other geographic factors. All climates on every continent exhibit such a dependence, but nowhere is it so simple, so direct, or so easily comprehensible as in North America. (1953: 125)

The central areas of the continent stretching from the Canadian Shield to the Gulf of Mexico, are, on the one hand, cut off from immediate Pacific influence by the high mountains stretching from Alaska virtually to the Mexican border. Meanwhile, they are open, unimpeded by any mountain ranges, to the contrasting influences of continental polar air from the north and maritime tropical air from the south, which helps explain the extraordinarily rapid fluctuations of temperature and the fact that it is here that tornadoes are more frequent than anywhere else in the world.

Complexity Theory and Geography

Complexity theory comprises a set of propositions for understanding how order in the world can emerge from disorder. It has no substantive content. Its claims can be applied regardless, which immediately suggests that it might be relevant to any project aimed at bringing physical and human geography together; hence, Massey's interest, though with the important proviso that human geography must necessarily be approached from a recognition of the crucial significance of the social – something that can be disregarded in physical geography to the extent that one can abstract from human intervention. Crucial to the emergence of some orderly arrangement out of a configuration of different parts are their nonlinear relations one with another. Force can generate a more – or less – than proportional effect: this can involve thresholds beyond which an effect suddenly increases or, in the other direction, suddenly decreases, though not necessarily. In this way, order can emerge from an apparently chance configuration of elements that come into relation with one another, that make connections, act as conditions for one another, perhaps. Hence order is something that is emergent and not reducible to its component parts, while those parts are modified in virtue of their relation to the whole. On the other hand, configuration is not necessarily a prelude to emergence; rather, it may simply be a matter of contingent effects where a condition remains exactly that and is not transformed into a necessary aspect of a broader system or structure.

Complex systems are self-organizing; order may emerge without any central control. Markets are a classic and obvious case, as in Adam Smith's "hidden hand," but so too is society. Equally clear are the storms studied by climatologists or the fractal geometries of drainage basins: the self-similarity of morphometric relations at different scales, both within the basin as a whole and within its tributary basins, which suggests that the order displayed in drainage basins might also be replicated in highway networks, as indeed Peter Haggett demonstrated (1967), or in central place systems.

Martin and Sunley (1996) have argued that spatial variability is fundamental to complexity since relative location affects interaction. This has been the basis for an understanding of the relation between so-called path dependence and development. And to be sure it is through the idea of path dependence – almost a cliché in human geography now, alas – that complexity theory first had an impact on human geography. It was borrowed from economics and then applied to a geographic context. But arguably it had put in an appearance much, much earlier, as in the work of Hägerstrand on migration (1957) and Curry (1966).

Complexity in Human Geography

We can consider this from the standpoint of two highly abstract processes in human geography: the concentrating and the dispersing. In both instances, chance configurations can, through the interaction of parts, generate these effects. Concentration assumes numerous and obvious forms: urbanization and agglomeration; regional specialization; the way in which migration streams from a particular place, particularly a smaller town, are focused on a very particular urban area (Shelley and Roseman 1978). To take one instance of how this might work, Peter Haggett (1965: 176) referred to the possibilities of understanding agricultural specialization as "scale concentration around a random nucleus," as for example, the rhubarb triangle in South Yorkshire. One can well imagine such a "random nucleus" emerging. Random spatial distributions show local clustering (see Figure 3.2). So by chance, some farmers in close proximity to one another, might find themselves producing the same specialty: spring flowers, soft fruit, rhubarb, or even potatoes. This nucleus then allows the breaching of certain thresholds: the point at which it becomes profitable to form a cooperative to take care of marketing, creating a regional brand, as in Idaho potatoes, and lobbying for various forms of state support.

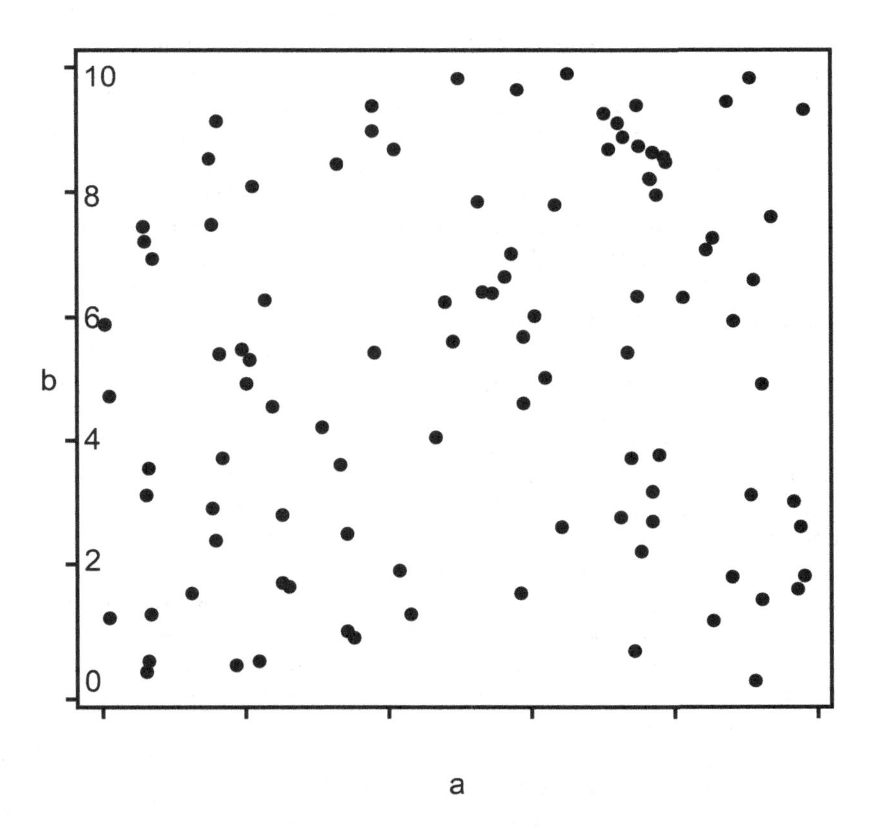

Figure 3.2 A random spatial distribution.

This conception of process has wider relevance. Exactly where and when gentrification occurs is often a puzzle. In larger cities there are often older houses of some architectural merit close to the historic downtown. Not all of them will be rehabilitated for middle class use. But given the nature of random distributions, it is always possible that those rehabbing without a view to gentrification as a wider process find themselves in close proximity, creating a neighborhood for each other, and attracting attention from others. One might then reasonably expect the formation of a residents' association to foment the process further: tours for banks to encourage lifting restrictions on mortgages for what was once regarded as a risky area; pressure on the city to revert to the old cobbled streets and installing street lamps of yesteryear. And so on. Further applications are only as extensive as one's imagination: the process can shed light on any form of path-dependent development, any emergent bias in the geography of uneven development broadly understood.[10,11] Hägerstrand's (1957) work on chain migration is classic.

This is to talk about concentrative effects: how a random distribution can, through the interaction of the parts, result in some sort of coming together. But it can also work in the other direction. The regularity in urban distributions is no accident but ultimately traceable to the search on the part of retailers for some market power: moving away from each other in order to find it. Those who make bad choices by locating "too close" to a competitor, simply go out of business. The same applies to the spacing of railroad lines or the hubs of airlines: a case of trying to maximize distance from others rather than minimizing it, as in concentrative effects. The spacing of regional shopping centers and the planning of developers is also clear.

This is perhaps a little too formulaic to do justice to how the idea of emergence applies to human geography. It is latent within any particular configuration of conditions and forces. Something like the energy-intensive US city only makes sense in terms of a coming together of a set of different conditions: the automobile and the oil industry, obviously, but also the real estate developers, and the mortgage lenders[12]; and, less appreciated, the highly fragmented jurisdictional pattern of American metropolitan areas, where any and every small burg on the urban periphery is keen for a share of the (taxable) loot. The "perfect storm" metaphor is a fitting one, and equally fitting, when shorn of its negative connotations, to the most concrete of formations like "the English countryside"[13]; and to the more abstract, like the emergence of the most recent round of globalization in the context of an equally disparate set of conditions and cumulative effects: the crisis of the late 70s, the monetarist turn, the container, the web, marketization in China, and lots more.

Geomorphology

An understanding of some aspects of landscape can be developed along similar lines. According to Harrison (2001: 328) "macroscopic landscapes are emergent phenomena." Summerfield also noted the possibilities in drawing parallels between the path-dependent development that is QWERTY and landscape development: "This example has obvious geomorphological parallels – for instance, in discordant drainage. In both cases an initial optimal adjustment to prevailing conditions creates a significant barrier to subsequent change – the investment barrier to retrain typists in one case, and the topographic barrier created by the superimposition of river valleys on to discordant structures in the other" (2005: 407).

This is exactly the case in the development of the English Lake District, as others (Dury 1959: 24) have pointed out. It has precisely the sort of discordant drainage system that Summerfield is referring to. Rivers cross boundaries between more and less resistant rocks regardless; there is no clear tendency for them to have exploited softer strata in carving out their channels (Figure 3.3). The rivers cross from the more resistant Borrowdale Volcanics to the less resistant slates to the north (Skiddavian) and to the south (Silurian).

Originally, where the Lake District is today, layers of sedimentary rocks ranging in age from the Carboniferous through the Permian to the Triassic had been lain down on ocean floors. These were superposed on much older rocks, more resistant to weathering and erosion, that are now on the surface: the Borrowdale Volcanics, the slates, and various igneous intrusions, which are particularly apparent in the west. Uplift created a dome-like structure with the Triassic on top. It was on this dome that the current pattern of rivers and their tributaries was laid down, which accounts for the radial pattern of the major valleys of the Lake District.

Just how the radial drainage pattern emerged is not hard to imagine. Any mild perturbation in the surface of the earth or variations in surface cover (Mosley 1972) would have channeled the water. Subsequent erosion and incision would have steepened the sides and called into play what Jonathan Phillips has called the principle of gradient selection,[14] which simply means that the initial gradients that have been selected in, quite by chance, endure and increase. Water and whatever load of sediment it has picked up takes the path of least resistance. In short, initial channels tend to persist and as the water, along with its burden of sediment ranging from the dissolved to bits of rock, erodes its channel, so incision increases. Similar processes would have been universal on the surface of the dome, the larger streams expanding their basins at the expense of neighbors through more rapid down-cutting and transport away of material moving down valley slopes.

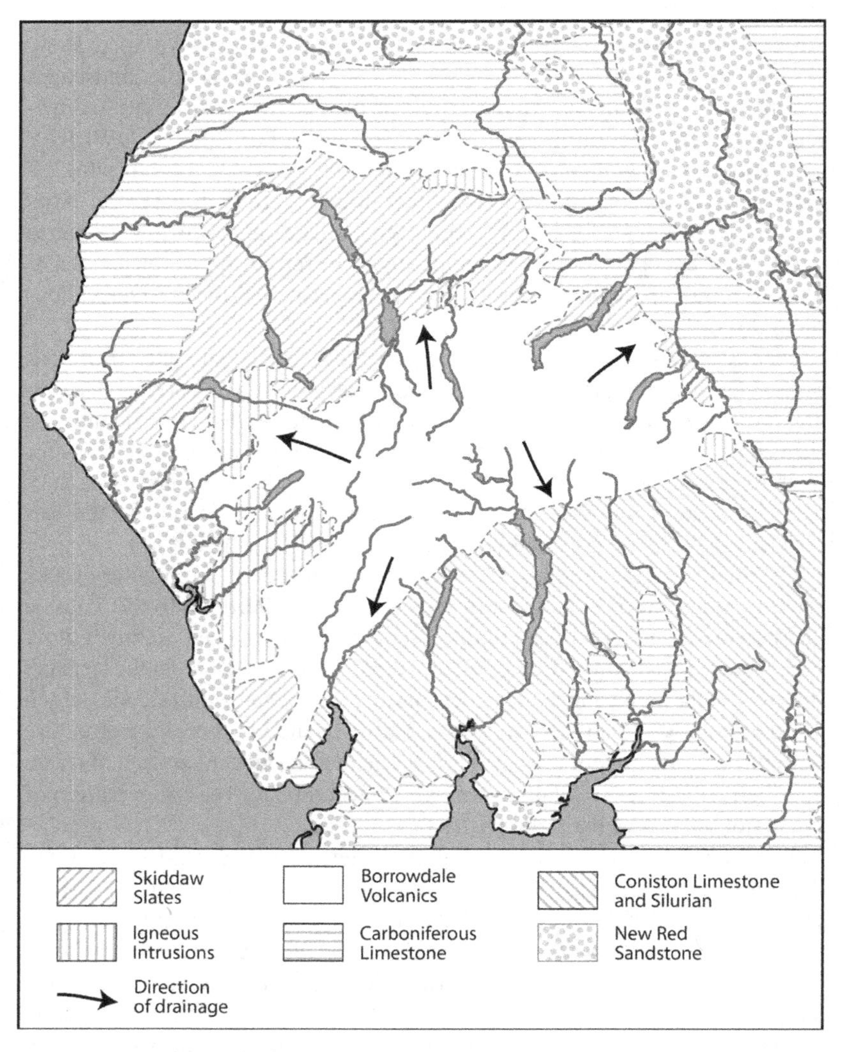

Figure 3.3 Geology of the English Lake District and the radiating drainage pattern.
Drawn by Jim DeGrand.

In the Lake District case, the overlying and weaker sedimentary rocks are long gone, eroded, and conveyed away by the streams that emerged on the dome: all that remains are now on the very fringes of the Lake District. But as the rivers encountered the more resistant rocks underneath – actually, as we have seen, rocks of *varying* resistance – the channels were already set and incised themselves further, hence the discordant drainage pattern evoked by Summerfield as a case of path-dependent development applied to the physical landscape. This would have been the topography later to be exposed to the effects of the ice age: snow collecting to a greater depth in the preexisting channels so that it would be there that glaciers would form and add to preexisting erosion by fluvial processes, erosion by the ice and its load of weathered away boulders and rocks of all sizes. The result is the highly glaciated landscape that we encounter today, including the lakes in basins eroded by the glaciers.

What are known as scarplands are common landscape features around the world: tilted blocks of strata of alternating resistance giving rise to a landscape of steep cuestas at the top of which the land slopes down gently in what is called a dip slope, the more resistant strata disappearing below the less resistant, to be then succeeded by another cuesta, and so on (see Figure 2.4). The ensemble of landforms thus constituted is then complemented by what are called outliers and water gaps. Water gaps are where rivers breach the more resistant strata of the cuestas or scarps and provide another example of discordant drainage with an origin similar to that of the English Lake District: rivers incised themselves into softer overlying strata and on reaching the more resistant ones underneath, existing gradient selection set in, allowing the drainage geometry to continue much as before; but then creating gaps in the scarps through which rivers flow. Outliers on the other hand, are the remnants of a retreating scarp: elevated areas with the same resistant cap-rock as the cuesta but now separated from it by softer strata at a lower elevation. There are numerous of these along the Jurassic scarp of the Cotswolds, as in the case of Bredon Hill and Oxenton Hill in Figure 3.4: both capped by the more resistant Birdlip Limestone, its resistance enhanced by its porosity. This

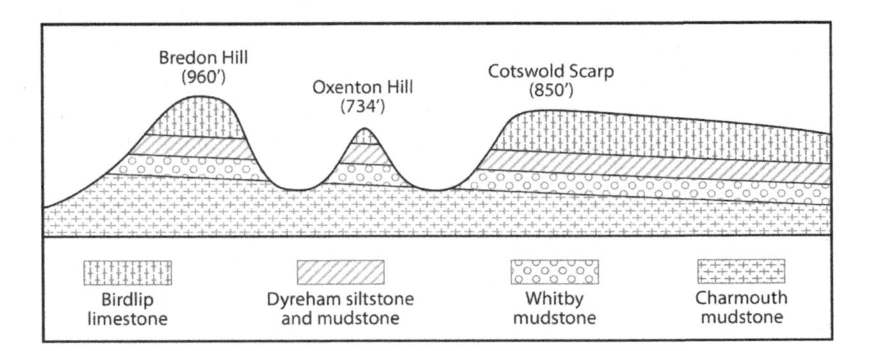

Figure 3.4 A schematic cross section of some outliers of the Cotswold escarpment.

Source: Drawn by Jim DeGrand.

is an example of what Jonathan Phillips calls "resistance selection," though it clearly goes hand in hand with his "gradient selection": on the retreating scarp face, one can imagine streams, each with their tributaries, incising laterally and thus isolating some portions as outliers-to-be. The tributaries of streams in water gaps would have the same effect.

Climatology

In geomorphology, memory of landforms past plays a huge part in understanding landscapes as emergent; the current drainage pattern of the English Lake District bears the clear traces of what happened millions of years ago. In climatology, this is much less evident. There *is* memory. The oceans store heat from the past and the fact that they lose it much more slowly than the land has an effect on climate. It can also happen, and by chance, that climate, and regional climate at that, can receive an inflection that can last for several centuries, as we will see in a discussion of the Little Ice Age. But to emphasize: typically centuries or less and not millions of years ago. This can also be compared with the social world. The Romans left an imprint on highway patterns still evident today. London and Paris have long been nationally dominant; an early size advantage had effects on transport networks, which then amplified their (relative) population growth.

It is standard teaching in introductory classes in climatology that climate is average weather. Certainly there are variations in the weather from one year to the next, but these tend to average out. Nevertheless that average can conceal some marked variations from year to year, which contribute to the distinctiveness of a particular climate. One knows that they will occur but not exactly when. Severe winters or particularly hot and dry summers are not the average in the British Isles but one can certainly expect them. Likewise, successions of exceptional weather from one year to the next: one unusually warm summer followed by another; or the series of drought years in the Great Plains that led to the Dust Bowl. How to make some sense of them?

From the physics of turbulent flow, which applies to both the atmosphere and the oceans which have such important implications for climate, we know that there are effects that do not necessarily cancel each other out and which can be mutually reinforcing. This on top of the randomness in the geography of climate memory stored in the oceans, not to mention the effects of topography and the turbulence generated on lee sides of mountain ranges lying astride storm paths, can create all manner of juxtapositions. These can have implications for such features of the weather as blocking highs or successions of storms, which, over time, because of their recurrence in some form or another and not necessarily in any predictable fashion but seemingly random over time and space, go to characterize a particular climate and its interannual variations. These random changes in underlying conditions can, through various positive feedback effects, have quite disproportionate implications for the climate. The temperature changes required to bring about the last ice age were, apparently, exceedingly modest (Bryson 1974).[15] The effects can also be regionally concentrated. Glaciation affected all of Northwest Europe and the Alpine area, but not Northeast Russia. In Northwest Europe there were copious snows allowing for the formation of glaciers. But in Northeast Russia, far from major storm belts, this was not the case.

Another case in point of regional effects is the Little Ice Age. This was a period of distinctly colder temperatures in Western Europe lasting from about 1300 into the nineteenth century. Alpine glaciers expanded, overrunning farms and villages, and the Baltic froze over. Iceland was particularly hard hit, losing half of its population. Exactly what caused it has long been uncertain. More recent analyses suggest that it started with a series of four volcanic eruptions in the tropics between 1250 and 1350. The resultant sulfurous dust in the atmosphere then impeded solar radiation reaching the earth's surface. The records show that cooling started between 1250 and 1300. Computer simulations then show how this could have resulted in a growth in the ice sheets in and around the Arctic Ocean. The Little Ice Age then received further impetus from a decline in sun spot activity starting around 1645, and this would have reduced in its own way solar radiation at the earth's surface. Resultant temperature changes in the North Atlantic would then have been reflected in what is known as the North Atlantic Oscillation: high pressure over the Azores and low pressure over Iceland – the positive phase – results in warmer than usual weather over Western Europe. But in its negative phase when the Azores high pressure is weaker, the North Atlantic storm tracks are pushed south, dragging in colder air behind them and into Western Europe.

Configuration, Contingency, and Complexity

I am proposing that one way of bringing physical and human geography together is through a particular sort of understanding of areal differentiation. This is in terms of configurations of conditions: those that occur in physical geography are always of a chance, circumstantial nature. Juxtaposition is central to understanding in human geography too, though some of that coming together is contrived in order to produce a useful effect: what occurs by chance is mobilized, given new meaning, put in new configurations, though contingent conditions always remain in the background. Developers try to maximize their rents through the artful arrangement of land uses in their developments, but their ultimate success also depends on conditions over which they have no control, like changes in the local labor market and hence in the demand for housing.

Something like the general circulation of the atmosphere can be regarded as the result of a particular intersection and interaction of different conditions and forces without reference to emergence. These conditions would include the fact that in virtue of the earth's sphericity, receipts of solar radiation vary by latitude generating differences in atmospheric pressure; the fact of the rotation of the earth then gives the movements of the air masses a directionality that deviates from perpendicularity to the isobars; the fact that the earth's axis slants gives us the seasons; the disposition of the various landmasses then generates variations in continentality producing the Asian monsoon, as well as in the patterns of movement of the ocean currents.

As we saw, chance events can then be the basis for nonlinear effects against which there are no compensating forces in the other direction. What was contingent to processes of socio- or physico-spatial formation becomes essential to it and to its reproduction. But it is not necessary that contingency should have those effects. Where it does, the emergent form itself can then become a condition for other changes of a contingent nature. The English Lake District is an emergent form, but it then created conditions for new sorts of processes and effects that are thoroughly contingent. Subsequent glaciation had nothing to do with the doming and the incision of streams down through softer sedimentary strata until they reached rocks of quite variable resistance underneath, creating the discordant drainage pattern discussed earlier. Glaciation was certainly structured by the pattern of radiating valleys but it had effects of its own: a new chapter in the history of the landscape. Most notably it created the characteristic U-shaped valleys and the pattern of lakes: lakes that are slowly being filled in by sediment deposited by streams from the adjacent slopes and to which quite extensively flat areas bear testimony.

A final example of the role of contingency and complexity and how they both have to be taken into account when explaining the role played by configuration is suggested by a book on Southeastern England. In *Rethinking the Region* (1988), John Allen, Doreen Massey, and Allan Cochrane set out to understand how Southeast England came to be a part of the way in which the regional geography of the United Kingdom is imagined. They dubbed it a neoliberal region and emphasized the chance coming together of a number of forces that gave it an economic élan. These included the deregulation of financial services in the 1980s, which gave the City of London such a boost, and the emergence of new growth poles of hi-tech industries around Cambridge and Oxford, and the axis of growth extending west of London and taking in Reading and Newbury. Subsequent growth pulled in the highly skilled from elsewhere in the United Kingdom and as the region became more and more important to government revenues, so it became the site for infrastructural projects designed to reduce the increasing congestion. In other words, the Southeast as emergent, though with other preconditions that go a long way back, most notably the concentration of government in London and a centralizing transport system focused on the capital city.

Configuration and Questions of Scale

> "How, if at all, can we hope to integrate the apparently incompatible results obtained from studies on such an enormous range of temporal and spatial scales?" . . . "beyond and above all these worries is the crucial preoccupation: can there be uniquely geomorphological theory, and if so, what will it look like?"
>
> (Kennedy 1977, 154).

Scale has always been a central concept in geography, even while the way it has been drawn on has had distinctly different histories across human geography, geomorphology, and climatology. It is, nevertheless, possible to identify certain recurring themes and issues: notably, and as discussed in Chapter 1, tendencies to see scales as self-sufficient, as defined by distinctive processes independent of those applying at other scales;[16] rather stereotypical, even rigid views of scale, like climatology's macro-, meso-, and micro: scale as absolute rather than relative – as in larger versus smaller; and then a focus on space rather than time-space. Drawing on ideas of configuration suggests a resolution of these recurring difficulties as well as bringing human and physical geography under the same conceptual umbrella.

Kennedy's concerns also apply beyond geomorphology to human geography and climatology. In human geography there are some notable exceptions to conventional treatments of scale: rather it is a conception where smaller scale configurations of forces form and then, in their turn help to transform, the larger scale. Harvey's geopolitics of capitalism and how it can be extended beyond the relation between countries and the global to take in that of urban areas is outstanding (1985a; 1985b). Storper and Walker's *The Capitalist Imperative* shows how the microscale of location theory can inform and be informed by the macroscale of uneven development. Capitalism on a global scale has changed its concrete form but always as a result of events at smaller scales. The most recent 'globalization' cannot be understood outside of a particular combination of events. These included the invention of the container, the newly industrializing countries of East Asia (NICs) and the crisis of accumulation in the developed countries. The latter then led firms there to outsource, taking advantage of the container and the NICs; which then gave capital on a global scale a new complexion. As Michael Storper (1987) remarked, "the local makes the global."

These are cases in which configurations can emerge in virtue of purposeful human intervention, though not necessarily. This cannot be the case in physical geography, where the spontaneity of forces rules.[17] Things come together, reflecting what is going on more globally, while creating the possibility of other configurations at smaller scales. Something like the basin-and-range landscapes of the Western United States have as their condition a particular combination of tectonic activity and location within the global circulation of the atmosphere. Their larger elements have then been the condition for smaller scale ones, most notably alluvial fans and pediments. Something like the scarplands of England, with their characteristic outliers and water gaps, can be interpreted in a similar way. The nesting of drainage basins of different order is a clear appeal to scale and its significance as formative and forming (DeBoer 1992). But quite aside from its limitation to fluvial landscapes, it needs to be situated with respect to the more scale extensive processes of tectonics and climate.

A similar idea of scale and scalar interrelations applies to the climates of the world. The zones along which midlatitude storms form are fairly, not completely, predictable, depending on the location of the different high-pressure belts. Where and when the storms form along these zones of convergence is predictable on average but no more than that; chance eddies in the atmosphere, perhaps reflecting differences in the temperature of the oceans at various points, and reinforcing one another to provide for more turbulent flow, can provide the nucleus.

An appreciation of just how connected places are and how the variation in weather that plays out there can be so variable has come from teleconnection research. The so-called El Nino is the best known of these teleconnections, creating changes in seasonal climate from the Horn of Africa all the way down to the Cape on the one hand, and down the east coast of the Pacific on the other. The North Atlantic oscillation referred to in the earlier discussion of the Little Ice Age is another: a see-saw of temperatures between Greenland and Iceland on the one hand and Western Europe on the other: so a bump in temperatures in the former, accompanied by cool summers and more severe winters in the latter, and vice versa. The changing map of atmospheric pressure that is the proximate condition for this, then alters the location of the storm belts. There is only one atmosphere and only one ocean suggesting that the basis of these anomalies is global. In corroboration of this point, a map showing the correlation between an index for the Southern Oscillation and sea-level atmospheric pressure is highly suggestive (Feldstein and Franzke 2017: 57) (see Figure 3.5). Moreover, there is evidence that what tips the balance from one to the other are extremely local events. According to Feldstein and Franzke (2017: 58), the generative mechanisms include local, self-reinforcing eddies at the tropical convergence zone between the two subtropical high-pressure belts and changes in the strength of the polar vortex. But regardless of the details, relatively local or regional changes in the circulation of the atmosphere are of crucial significance.

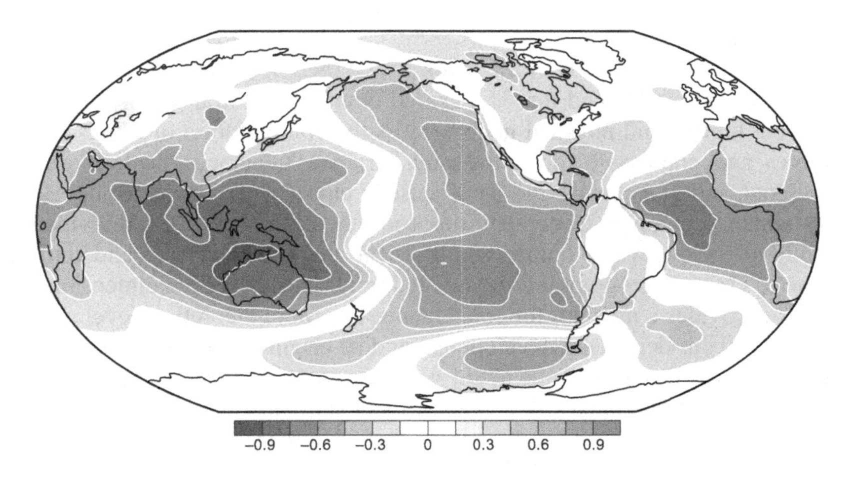

Figure 3.5 The correlation between a measure of the intensity of the El Nino effect and of sea-level pressure. The intensity of the El Nino effect is measured as the difference in sea-level pressure between Darwin, Australia, and Tahiti.

Source: After Feldstein and Franzke 2017: Figure 3.3, p. 57.

Meanwhile, the global circulation has to be conceptualized as a more complex configuration of forces and conditions than might be suggested by conventional notions of the different climates and their belt-like character. Differences in net radiation over the earth surface crucially combine with the distribution of land and water and the geography of elevation above sea level to deform what might emerge over a more homogeneous earth. The fact of the Asian monsoon is one aspect of this. The fact that most of the world's land is in the northern hemisphere then affects winter temperature gradients in a way that is not replicated in the southern hemisphere. Major high elevation complexes of mountains and plateaus exercise deflecting effects on the characteristic westerly flow of air with further consequences for the differentiation of climate across the surface of the earth (Saulière et al. 2012).

******

Academic geography has a problem of fragmentation. It comprises two branches that share very little in their research or what they teach. They seem to lack a commonality that might encourage interaction. In Chapter 1, some approaches to countering this were discussed, all evident in the literature, but all of them found wanting in some respect or another. Just why it is so difficult was then outlined in Chapter 2. For the most part, people now seem to have given up, aside from a minority of mainly geomorphologists and British ones at that – a curiosity in itself. This is a pity. It betrays the promise of the broader question that got many people into the field in the beginning: simply put, why is it different here compared with other places that one is aware of?

What I have suggested in this book is that there is something quite crucial that they share, and that is a spatial ontology. Central to explanation in both human and physical geography is the configuration of things and forces, and, one should add, at the same variety of geographic scales: local, regional, and global make the same sense in physical geography as they do in human geography. Concretely this could be at the heart of a revivified study of regions where human and physical processes converge and combine to create a geography differentiated in all its substantive dimensions. It also provides a medium for convergence at a more theoretical level. The notion of path-dependent development and its origins in some chance configuration applies to both and could stimulate cross-fertilization.

Finally, and returning to the differences of substance that have tended to keep the two subfields apart, there is the fact that each can provide an important counterpoint function, making the other more self-conscious of its particular biases, obsessions, and oversights. I suggested that we can learn a lot about human geography's obsession with space by looking at physical geography's indifference and why that might be, even while it is as spatial in its understandings as human geography. But it works the other way around too: how can it be (physical) geography if it *doesn't* obsess about space?

Notes

1 Compare Brierley et al.: "Recognizing that every location is potentially unique does not render generalizations meaningless. Regularities in time and space can still be observed as repeated patterns of landforms, and interpretations of these patterns can support efforts to meaningfully transfer understandings from one location to another. The challenge lies in identifying where a general pattern holds true and how and when local differences may be important" (2013: 602). These views are echoed by Favis-Mortlock and De Boer: "Returning to the question that began this chapter: perhaps it is time for geomorphology to mature as a science. Perhaps it is time to recognize that, apart from idle curiosity, explanation of specific features in the landscape is, by itself, of limited value. Where such explanations are essential is in validating our general models of landscape development. Thus it is of limited interest to know why the sediment load of the Rhine River varies as it does; but it is a different matter if knowledge of the Rhine's sediment-carry characteristics helps us to understand the unifying principles controlling the sediment loads of the River Exe, the Rhine River and the Amazon River. Similarly, we do not really need to explain the details of the drainage pattern of the South Saskatchewan River. Instead, we should look for similarities between, and common causes of, the drainage patterns of this river and other rivers at various scales in different regions." (Favis-Mortlock and De Boer 2003); available here: www.researchgate.net/publication/318542568_Simple_at_heart_Landscape_as_a_self-organizing_complex_system (last accessed August 20, 2022.)

2 Paul Claval has pointed out to me how in mid-century French geography the idea of "combinaison" or combination was seen as the distinctive contribution of the discipline. He has referred me to the second edition of André Cholley's (1951), *La Géographie, Guide de l'étudiant*. Thus: « A la différence de la plupart des sciences objectives qui cherchent à isoler les faits qu'elles observent . . . la géographie considère la réalité dans sa complexité même. » And: « Le fait géographique, même le plus simple, exprime toujours une combinaison, une convergence d'éléments et de facteurs divers. C'est son essence même:" English translation : In contrast to most of the sciences which seek to isolate the facts that they observe . . . geography considers reality in its complexity. The geographic fact, even the most simple, always expresses a combination, a convergence of elements and diverse factors. It is its essence."

3 With regard to climatology Phillips has echoed these sentiments: "In many situations, the best way to forecast the weather is not via cause-effect relationships or global principles of atmospheric physics, but based on a typology of situations that recognizes and incorporates the unique characteristics of places. This is the foundation of synoptic meteorology and climatology, which is in turn the foundation of weather and climate forecasting. Synoptic climatologies also are concerned explicitly with time, at least in that they are sensitive both to seasonality, and to sequences of events" (2001: 9). In his paper on "The perfect landscape," Phillips also explores the significance of configuration in geomorphology and its relation to the immanent, though with some tendency to reduce configuration to simple contingency and not follow through to emergence. Thus: "Landscapes are indeed shaped and controlled by deterministic, global laws, but the operation of these laws in specific geographical and historical contexts means that landforms and landscapes are often circumstantial, contingent outcomes, not derivable from global laws alone" (2007: 167).

4 To some degree, Les Curry had anticipated her, both by emphasizing the role of chance processes and seeing the applicability to both human and physical geography: "To conceive the landscape as composed of elements having characteristic spectra or fluctuations in time and of differentiation in space, to depict society as an organization for facing the future and choosing among a variety of possibilities, to stress movement and development, to place uncertainty in the center of our analysis instead of ignoring it as regrettable, all this must be welcome. That the format of the language of probabilities is equally germane to discussing both physical and human events, both spatial and temporal ordering. both retrospective and anticipatory explanations of actions, both individual and collective activities, the physical world both as exclusive of man and as the environment of man, bespeaks its versatility. A common mode of discourse thus presents itself for scientific study of the landscape, promising to promote an articulated coherent viewpoint" (1966: 54). Another important reference is Stoddart (1966) who, in an article on Darwin and geography, underlined the significance of his claims about chance variation and the selection in of certain traits.

5 This is an approach evident in Allen, Massey and Cochrane (1998).

6 "The philosophy of realism implies a structuring of the world in which observable events are contingent on appropriate circumstances for mechanisms to act. Methodologically, this means that interpretation of the nature of mechanisms from observation of events demands an understanding of the role played by local conditions in time and space" (Richards 1996: 182).

7 "Perspectives of interactions over time have moved beyond the simple sense implied by Davisian evolutionary models, recognizing that the history of change and perturbation in landscapes is a key determinant of subsequent geomorphic activity. At the same time, geomorphologists have embraced space in new and different ways, emphasizing the importance of spatial arrangement – the way that landscapes are configured and their topology" (Preston et al. 2011: 30).

8 The contrast here is with work on the totality of landscape as in geomorphometry or even the earlier morphometric work aimed at summarizing and contrasting landscapes, as in Chorley and Morgan (1962).

9 Compare an early attempt to address the regional specificity of climate: Trewartha's *The Earth's Problem Climates* (see also Dixon and Herbert 2018).

10 Suburbanization around major cities is a very uneven, directionally-biased process. New development in a particular sector takes off for no obvious reason, gathers momentum, is fed by new public infrastructure, attracting in still newer development.

11 It also works in the reverse direction. The city of San Bernardino experienced a series of quite unrelated setbacks to employment in a relatively short space of time: the closure of a steel mill at nearby Fontana (1983), the closure of a US air force base (1994) that had been a major employer, and the closure of a big marshalling yard for the Santa Fe railroad (1992). Declining real estate values then made it attractive to adjacent cities, most notably Los Angeles, as a dumping ground for Section 8 recipients, lending further impetus to the downward spiral: a sort of reverse path dependence.

12 See Walker and Large (1974) and, in particular, their argument about spiral development.

13 https://kevinrcox.wordpress.com/2018/10/03/landscapes-urban-geographies-and-how-england-is-different/

14 "Mass and energy fluxes occur along the steepest gradients of potentials or concentrations. The principle of gradient selection is simply that geomorphic features associated with these gradients persist and grow. For instance, hydraulic selection principles operate as water is redistributed on the surface. The most efficient paths persist and prevail; less efficient options dry up. As channels are scoured due to initially chance concentrations of shear stress relative to local resistance, hydraulic selection favors channels over diffuse flow, and the steepest, largest, and smoothest channels over smaller, flatter and rougher ones" (Phillips 2011: 320–321).

15 "Most careful analyses suggest a mean global surface temperature difference from full glacial to the present of 4°C to 6°C, but with smaller and larger changes in specific areas. Most of the changes deduced from the geologic record, for specific places, fall within the range of individual monthly anomalies for the historic record" (Bryson 1974: 753).

16 As in the much cited paper by Schumm and Lichty.

17 Though one should note how configuration in physical geography can be the inadvertent result of human activity whose purpose was otherwise: the implications of reservoir construction for stream flow and erosion, as well as upstream base levels; or of the pumping of ground water for subsidence would be examples.

References

Alexander J (1963) *Economic Geography*. New York: Prentice Hall.

Allen J, Massey D and Cochrane A (1998) *Rethinking the Region: Spaces of Neo-Liberalism*. London: Routledge.

Anderson MS (1951) *The Geography of Living Things*. London: England Universities Press.

Barry RG (1970) A framework for climatological research with particular reference to scale concepts. *Transactions of the Institute of British Geographers* 49, 61–70.

Barry RG and Chorley RJ (2003) *Atmosphere, Weather and Climate*. London: Routledge.

Beckinsale R and Beckinsale M (1980) *The English Heartland*. London: Duckworth.

Beresford M (1954) *The Lost Villages of England*. London: Lutterworth Press.

Berry BJL (1967) *The Geography of Market Centers and Retail Distribution*. Englewood Cliffs, NJ: Prentice Hall.

Bond S and Featherstone D (2009) The possibilities of a politics of place beyond place? A conversation with Doreen Massey. *Scottish Geographical Journal* 125, 401–420.

Borchert JR (1950) The climate of the central North American grasslands. *Annals of the Association of American Geographers* 40:1, 1–49.

Boyce RR and Clark WAV (1964) The concept of shape in geography. *Geographical Review* 54:4, 561–572.

Brierley G, Fryirs K, Callu MC, Tadaki M, Huang HQ and Blue B (2013) Reading the landscape: integrating the theory and practice of geomorphology to develop place-based understandings of river systems. *Progress in Physical Geography* 37:5, 6–1–621.

Bryson R (1974) A perspective on climate change. *Science* 184, 753–760.

Bunge W (1962) *Theoretical Geography* (Lund Studies in Geography, Series C, General and Mathematical Geography 1).

Chorley RJ and Haggett P (1960) *Network Analysis in Geography*. London: Edward Arnold.

Chorley RJ, Malm DEG and Pogorzelski HA (1957) A new standard for estimating drainage basin shape. *American Journal of Science* 255, 138–141.

Chorley RJ and Morgan MA (1962) Comparison of morphometric features, Unaka Mountains, Tennessee and North Carolina, and Dartmoor, England. *Geological Society of America Bulletin* 73, 17–34.

Clifford N (2001) Editorial: physical geography – the naughty world revisited. *Transactions of the Institute of British Geographers* NS 26, 387–389.

Clifford NJ (2002) The future of geography: when the whole is less than the sum of its parts. *Geoforum* 33, 431–436.

Cosgrove D (2006) Modernity, community, and the landscape idea. *Journal of Material Culture* 11:1–2, 49–66.

Cox KR (2013) Notes on a brief encounter: critical realism, historical materialism and human geography. *Dialogues in Human Geography* 3, 3–21.

Cox KR (2014) *Making Human Geography*. New York: Guilford Press.

Cox KR (2016) Geographies, critical and Marxist, and lessons from South Africa. *Human Geography* 9:3, 10–26.

Cronon W (1991) *Nature's Metropolis*. New York: WW Norton and Co.

Curry L (1966) Chance and landscape. In JW House (ed.) *Northern Geographical Essays*. Newcastle-upon-Tyne: Department of Geography, University of Newcastle-upon-Tyne.

Curry L (1972) A spatial analysis of gravity flows. *Regional Studies* 6, 131–147.

Darby HC (1951) The changing English landscape. *Geographical Journal* 117:4, 377–394.

Das BC, Islam A and Sarkar B (2022) Drainage basin shape indices to understanding channel hydraulics. *Water Resources Management* 36, 2523–2547.

Davis M (1995) The case for letting Malibu burn. *Environmental History Review* 19:2, 1–36.

DeBoer DH (1992) Hierarchies and spatial scale in process geomorphology: a review. *Geomorphology* 4, 303–318.

Dixon DP, Hawkins H and Straughan ER (2013) Wonder-full geomorphology: sublime aesthetics and the place of art. *Progress in Physical Geography* 37:2, 227–247.

Dixon RW and Herbert J (2018) Glenn Trewartha's *The Earth's Problem Climates* (1961.) *Progress in Physical Geography* 42, 128–133.

Dobson J (1992) Spatial logic in paleogeography and the explanation of continental drift. *Annals, Association of American Geographers* 82, 187–206.

Dury GH (1959) *The Face of the Earth*. Harmondsworth: Penguin Books.

Favis-Mortlock D and DeBoer D (2003) Self-organisation and complexity: a new perspective on landscape dynamics. In A Roy and ST Trudgill (eds.) *Contemporary Meanings in Physical Geography*. London: Routledge.

Fay B (1975) *Social Theory and Political Practice*. London: Allen and Unwin.

Feldstein SB and Franzke CLE (2017) Atmospheric teleconnection patterns. In *Nonlinear and Stochastic Climate Dynamics*. Cambridge: Cambridge University Press.

Fotheringham AS (1981) Spatial structure and distance-decay parameters. *Annals of the Association of American Geographers* 71, 425–436.

Giddens A (1984) *The Constitution of Society: Outline of the Theory of Structuration.* Cambridge: Polity Press

Goudie A (1981) *The Human Impact on the Natural Environment: Past, Present and Future.* Oxford: Blackwell.

Goudie A (2002) Aesthetics and relevance in geomorphological research. *Geomorphology* 47, 245–249.

Goudie A (2017) The integration of human and physical geography revisited. *Canadian Geographer* 61, 19–27.

Goudie A and Viles H (2010) *Landscapes and Geomorphology: A Very Short Introduction.* Oxford: Oxford University Press.

Gould PR (1975) Acquiring spatial information. *Economic Geography* 51, 87–99.

Graf WL (1979) Mining and channel response. *Annals of the Association of American Geographers* 69, 262–275.

Grove AT (1969) *Africa South of the Sahara.* Oxford: Oxford University Press.

Grove AT and Rackham O (2003) *The Nature of Mediterranean Europe: An Ecological History.* New Haven, CT: Yale University Press.

Hägerstrand T (1957) Migration and area: survey and sample of Swedish migration fields and hypothetical considerations on their genesis. *Lund Studies in Geography, B, Human Geography* 13, 27–158.

Hägerstrand T (1984) Presence and absence: a look at conceptual choices and bodily necessities. *Regional Studies* 18:5, 373–379.

Haggett P (1965) *Locational Analysis in Human Geography.* London: Edward Arnold.

Haggett P (1967) An extension of the Horton combinatorial model to regional highway networks. *Journal of Regional Science* 2, 281–290.

Haggett P (1990) *The Geographer's Art.* Oxford: Blackwell.

Haggett P (2012) *The Quantocks.* Chew Magna., Somerset: The Point Walter Press.

Hare FK (1953) *The Restless Atmosphere.* London: Hutchinson's University Library.

Harrison S (2001) On reductionism and emergence in geomorphology. *Transactions of the Institute of British Geographers* NS 26, 327–339.

Harvey D (1973) *Social Justice and the City.* London: Edward Arnold.

Harvey D (1985a) The Geopolitics of capitalism. In D Gregory and J Urry (eds.) *Social Relations and Spatial Structures.* London: Macmillan.

Harvey D (1985b) The place of urban politics in the geography of uneven capitalist development. In *The Urbanization of Capital.* Baltimore, MD: Johns Hopkins University Press.

Harvey D (1987) Three myths in search of a reality in urban studies. *Environment and Planning A: Society and Space* 5, 367–376.

Harvey D (1989) *The Condition of Postmodernity.* Oxford: Basil Blackwell.

Harvey D (1996) *Justice, Nature and the Geography of Difference.* Oxford: Blackwell.

Harvey D (2006) Space as a keyword. In N Castrate and D Gregory (eds.) *David Harvey.* Oxford: Blackwell.

Harvey D (2008) On the deep relevance of a certain footnote in Marx's Capital. *Human Geography* 1, 26–31.

Harvey D (2010) *The Enigma of Capital.* London: Profile Books.

Herbertson AJ (1905) The major natural regions: an essay in systematic geography. *Geographical Journal* 25, 300–310.

Hewitt K (1983) The idea of calamity in a technocratic age. In K Hewitt (ed.) *Interpretations of Calamity* London: Allen and Unwin.

Hoskins WG (1955) *The Making of the English Landscape*. London: Hodder and Stoughton.

Huntington E (1915) *Civilization and Climate*. New Haven, CT: Yale University Press.

Johnston RJ (1975) Map pattern and friction of distance parameters: a comment. *Regional Studies* 9, 281–283.

Jones JP and Hanham RQ (1995) Contingency, realism and the expansion method. *Geographical Analysis* 27, 185–207.

Kendrew WG (1922) *The Climates of the Continents*. Oxford: Clarendon Press.

Kennedy B (1977) A question of scale. *Progress in Physical Geography* 1, 154–157.

Kennedy B (1979) A naughty world. *Transactions of the Institute of British Geographer NS* 4, 550–558.

Kennedy B (2006) *Inventing the Earth*. Oxford: Blackwell.

Lambert JM, Jennings JM, Smith CT and Green C (1960) *The Making of the Broads: A Reconsideration of their Origin in the Light of New Evidence*. London: The Royal Geographical Society.

Lave R et al. (2014) Intervention: critical physical geography. *Canadian Geographer* 58, 1–10.

Leopold L, Wolman MG and Miller JP (1964) *Fluvial Processes in Geomorphology*. San Francisco, CA: W H Freeman.

Lubbock J (1902) *The Scenery of England*. London: Macmillan.

Mackinder HJ (1907) *Britain and the British Seas*. Oxford: Clarendon Press.

Malanson GP, Scuderi L, Moser KA, Willmott CJ, Resler LM, Warner TA and Mearns LO (2014) The composite nature of physical geography: moving from linkages to integration. *Progress in Physical Geography* 38, 3–18.

Manley G (1952) *Climate and the British Scene*. London: Collins.

Marr J (1900) *The Scientific Study of Scenery*. London: Methuen.

Martin R and Sunley P (1996) Complexity thinking and evolutionary economic geography. *Journal of Economic Geography* 7:573–601.

Martin Y and Church M (2004) Numerical modelling of landscape evolution: geomorphological perspectives. *Progress in Physical Geography* 28, 317–339

Marx K (1857–58; reprinted 1973) *Grundrisse*. Harmondsworth: Penguin Books.

Massey D (1999a) Spaces of politics. In D Massey, J Allen and P Sarre (eds.) *Human Geography Today*. Cambridge: Polity Press.

Massey D (1999b) Space-time, 'science' and the relationship between physical geography and human geography. *Transactions of the Institute of British Geographers* 24, 261–276.

Massey D (2005) *For Space*. London: Sage.

McDowell L (1997) *Capital Culture: Gender at Work in the City*. Oxford: Blackwell.

Miller AA (1953) *The Skin of the Earth*. London: Methuen.

Mitchell D (1996) *The Lie of the Land*. Minneapolis, MN: University of Minnesota Press.

Mitchell JB (1962) *Great Britain: Geographical Essays*. Cambridge: Cambridge University Press.

Mosley MP (1972) Evolution of a discontinuous gully system. *Annals of the Association of American Geographers* 62:4, 655–663.

Nyestuen JD (1963) Identification of some fundamental spatial concepts. *Papers of the Michigan Academy of Science, Arts and Letters* 48, 373–84 [Reprinted in BJL Berry and DF Marble (eds.) *Spatial Analysis* Englewood Cliffs, NJ: Prentice Hall, 35–41].

Phillips JD (2001) Human impacts on the environment: unpredictability and the primacy of place. *Physical Geography* 22, 321–332.

Phillips JD (2007) The perfect landscape. *Geomorphology* 84, 159–169.

Phillips JD (2011) Emergence and pseudo-equilibrium in geomorphology. *Geomorphology* 132, 319–326.

Preston N, Brierly G and Fryirs K (2011) The geographic basis of geomorphic enquiry. *Geography Compass* 5:1, 21–34

Rayner JN (1971) *An Introduction to Spectral Analysis*. London: Pion.

Rayner JN and Golledge RG (1972) Spectral analysis of settlement patterns in diverse physical and economic environments. *Environment and Planning A* 4, 347–371.

Rhoads B (1994) On being a 'real' geomorphologist. *Earth Surface Processes and Landforms* 19, 269–272.

Richards K (1996) Samples and cases: generalization and explanation in physical geography. Chapter 7 in BL Rhoads and CE Thorn (eds.) *The Scientific Nature of Geomorphology: Proceedings of the 27th Binghamton Symposium in Geomorphology*. New York: John Wiley.

Rushton G (1969) Analysis of spatial behavior by revealed preference. *Annals of the Association of American Geographers* 59, 391–400.

Sack RD (1972) Geography, geometry and explanation. *Annals of the Association of American Geographers* 62, 61–78.

Saulière J, Brayshaw DJ, Hoskins B and Blackburn M (2012) Further investigation of the impact of idealized continents and SST distributions on the northern hemisphere storm tracks. *Journal of the Atmospheric Sciences* 69, 840–856.

Sayer A (1985) The difference that space makes. In D Gregory and J Urry (eds.) *Social Relations and Spatial Structures*. London: Macmillan.

Sayer A (1986) New developments in manufacturing: the just-in-time system. *Capital and Class* 10, 43–72.

Sayer A (1989) Postfordism in question. *International Journal of Urban and Regional Research* 13, 666–696.

Sayer A (1991) Beyond the locality debate: deconstructing geography's dualisms. *Environment and Planning* A, 23:2, 283–308.

Sayer A (1992) *Method in Social Science: A Realist View*. London and New York: Routledge.

Sayre N (2010) Climate change, scale, and devaluation: the challenge of our built environment. *Washington and Lee Journal of Energy, Climate, and the Environment* 1, 93–105.

Schumm S and Lichty RW (1965) Time, space, and causality in geomorphology. *American Journal of Science* 263, 110–119.

Scott AJ (2014) *Solway Country: Land, Life and Livelihood in the Western Border Region of England and Scotland.* Cambridge: Cambridge Scholars Publishing.

Shelley FM and Roseman CC (1978) Migration patterns leading to population change in the nonmetropolitan South. *Growth and Change* 9:2, 14–23.

Simpson G (1963) Historical science. In CC Albritton (ed.) *The Fabric of Geology*. Reading, MA: Addison-Wesley.

Slaymaker O (2009) The future of geomorphology. *Geography Compass* 3:1, 329–349.

Soja EW (1985) The spatiality of social life: towards a transformative retheorization. In D Gregory and J Urry (eds.) *Social Relations and Spatial Structures*. London: Macmillan.

Sparks BW (1979) *Rocks and Relief.* London: Longman.

Stamp D (1946) *Britain's Structure and Scenery*. London: Collins.

Stoddart DR (1966) Darwin's impact on geography. *Annals of the Association of American Geographers* 56, 683–698.

Storper M (1987) The post-Enlightenment challenge to Marxist urban studies. *Environment and Planning A: Society and Space* 5:4, 418–426.

Storper M and Walker RA (1989) *The Capitalist Imperative*. Oxford: Blackwell.

Strahler A (1980) Systems theory in physical geography. *Physical Geography* 1, 1–27.

Summerfield MA (2005) A tale of two scales, or the two geomorphologies. *Transactions of the Institute of British Geographers* NS 30, 402–415.

Summerfield MA (2014) *Global Geomorphology*. London: Routledge.

Swyngedouw E (2007) Technonatural revolutions: the scalar politics of Franco's hydro-social dream for Spain, 1939–1975. *Transactions of the Institute of British Geographers* NS 32, 9–28.

Taylor PJ (1971) Distances within shapes: an introduction to a family of finite frequency distributions. *Geografiska Annaler* 53B, 40–53.

Taylor PJ (1981) A materialist framework for political geography. *Transactions of the Institute of British Geographers* NS 7, 15–34.

Thomas WL (1956) *Man's Role in Changing the Face of the Earth.* Chicago, IL: University of Chicago Press.

Thrift NJ (1983) On the determination of action in space and time. *Environment and Planning D: Society and Space* 1, 23–57.

Trewartha G (1954) *An Introduction to Climate*. New York, NY: McGraw-Hill.

Trewartha G (1961) *The Earth's Problem Climates*. Madison, WI: University of Wisconsin Press.

Trueman A (1949) *Geology and Scenery in England and Wales*. Harmondsworth: Penguin Books.

Tuan Y-F (1959) *Pediments in Southeastern Arizona*. University of California Publications in Geography 13.

Walker RA and Large DB (1975) The economics of energy extravagance. *Ecology Law Quarterly* 4, 963–985.

Ward R (2018) *Climate*. New York, NY: AMS Press.
Watts MJ (1983) Hazards and crises: a political economy of drought and famine in Northern Nigeria. *Antipode* 15, 24–34.
Wilkinson BH and McElroy BJ (2007) The impact of humans on continental erosion and sedimentation. *Geological Society of America Bulletin* 119, 140–156.
Woldenberg MJ (1969) Spatial order in fluvial systems: Horton's laws derived from mixed hexagonal hierarchies of drainage basin areas. *Geological Society of America Bulletin* 80, 97–112.
Wooldridge SW and Goldring F (1953) *The Weald*. London: Collins.

Index

Printed in the United States
by Baker & Taylor Publisher Services